D. Hellwinkel
Systematic Nomenclature of Organic Chemistry

Springer

Berlin
Heidelberg
New York
Barcelona
Hong Kong
London
Milan
Paris
Singapore
Tokyo

D. Hellwinkel

Systematic Nomenclature of Organic Chemistry

A Directory to Comprehension and Application of its Basic Principles

With 35 Tables

 Springer

Professor Dr. D. Hellwinkel
Organisch-Chemisches Institut
Universität Heidelberg
Im Neuenheimer Feld 270
69121 Heidelberg
Germany

ISBN 3-540-41138-0 Springer-Verlag Berlin Heidelberg New York

Cataloging-in-Publication Data applied for
Die Deutsche Bibliothek – CIP-Einheitsaufnahme
Hellwinkel, Dieter:
Systematic nomenclature of organic chemistry : a directory to
comprehension and application of its basic principles ; with 35 tables
/ D. Hellwinkel. – Berlin ; Heidelberg ; New York ; Barcelona ;
Hongkong ; London ; Milan ; Paris ; Singapore ; Tokyo : Springer, 2001
ISBN 3-540-41138-0

Springer-Verlag Berlin Heidelberg New York
a member of BertelsmannSpringer Science+Business Media GmbH

http://www.springer.de

© Springer-Verlag Berlin Heidelberg 2001
Printed in Germany

Typesetting: Fotosatz-Service Köhler GmbH, Würzburg
Coverdesign: design & production, Heidelberg

Printed on acid-free paper   SPIN: 10749737   2/3020mh – 5 4 3 2 1 0

Preface

The explosive growth of chemical knowledge during the last decennia has led to a sheer unbounded number of new and novel compounds and compound classes whose rational naming has caused ever-increasing difficulties. Originally the naming of a new substance was left exclusively to the discretion of the discoverer, who frequently derived the name directly from a sensory perception or even acted purely by intuition. As such more or less arbitrarily formed "trivial names" conveyed nothing about the structure of the underlying compounds, no reasonable chemical correlations could be established with those names. As a corollary of the deepening comprehension of structural relationships in Organic Chemistry it eventually became unavoidable to develop a binding systematic and universally applicable nomenclature framework permitting the encoding of essential constitutional information in a compound name proper.

One of the main difficulties in realizing such a task is to be found in the seemingly unresolvable conflict between the necessary rigidity of a system of norms and a certain degree of flexibility required in its application in teaching texts, handbooks, and secondary and primary literature. As a matter of fact, however, this situation has improved considerably in recent times insofar as the master concepts for obtaining rational, systematic compound names are now defined so precisely and uniformly that they are generally applicable and extendable.

In view of an ever-growing number of new chemical compounds it has become inevitable to adhere more and more to a stringently rational nomenclature system based on easily comprehensible, preferably self-evident, maxims instead of perpetuating the multiplicity of traditional naming procedures frequently overloaded with only vaguely justified exceptions. Accordingly, these guidelines for the application of the basic principles of systematic nomenclature emphasize above all fully systematic names corresponding to farthest reaching standardisation; this holds even when the official recommendations of the IUPAC nomenclature commissions have, an occasion, stopped half-way to their goal. Since in practice it will be impossible to jettison, at least for some time, a large number of widely used and accepted traditional and/or trivial names, these are also duly accounted for as far as possible. To aid understanding, explanatory comments are occasionally interspersed.

Finally it should be noted that nowadays most of the not too complex naming problems can be solved – quasi blindly – making use of an appropriate computer program. Notwithstanding this, it is the firm conviction of the author that some elementary insight into function and workability of chemical language belongs to the basic intellectual outfit of a chemist. The main aim of this introductory treatise is, therefore, to show chemists of every kind and level how to cope with their predominant means of communication, i.e., systematic chemical nomenclature. After all are we dealing here with a meticulously elaborated, strictly rational artificial language wich reflects in its organisation the characteristic analytical and synthetical thought and argumentation patterns proper to chemistry. And, moreover, even from a purely linguistic viewpoint such an endeavour is surely not without a certain charm.

December 2000   *D. Hellwinkel*

Contents

Introduction

The downright indifference or even aversion of many chemists to appela-
tion problems of their science is to some degree understandable since
there are simply too many divergent and inhomogeneous nomenclature
systems to choose from. Moreover, the same naming principles are frequently applied quite differently by different chemical journals, textbooks
and handbooks. Meanwhile, however, consensus has been reached to
use, wherever possible, an internationally binding uniform nomenclature
system that nevertheless tolerates certain alternatives. It has therefore
become necessary for every chemist – whether student or professionally
active – to acquire at least an elementary working knowledge of this system. This is all the more urgent since indexing by the globally active
Chemical Abstracts Service (Chem. Abstr.) and by Beilsteins Handbook of
Organic Chemistry is generally based on the IUPAC nomenclature rules,
even though certain deviations or extrapolations therefrom are often
employed. This will be taken into account where appropriate.

The unease generated by the rules of systematic nomenclature can
probably be dispelled somewhat by the following remarks: the naming of
a chemical compound and the derivation of a structure from a given systematic name follows the same general principles as chemical synthesis
and constitutional determination by degradation, respectively. The chemical structure to be named is broken down into its constituents which are
then given the appropriate systematic designations. The name fragments
thus obtained are then combined to the full name according to a definite
set of rules. On the other hand, a compound name is translated into the corresponding structural formula by separating it into its nomenclatural
subunits which are assigned partial formulae that then are joined together
to give the full structural representation.

The basic idea of systematic nomenclature, whose various modes of
application are to be conveyed here, resides in the concept of the "parent
structure" – an acyclic or (poly)cyclic hydrocarbon or hetero system –
whose hydrogens can be substituted by other atoms, groups of atoms,
or even subordinate parent systems themselves. These substituents can
likewise be further substituted in various ways. At the same time there

exists, particularly in the domain of parent structures, a plethora of trivial names thought to be indispensable for various reasons. These trivial names usually tell us nothing about the constitution of the compounds they represent, nor can they be derived logically. However, as they are often the starting points of a whole nomenclatural subsystem (e. g.: the fused polycycles), we simply have to memorise as large a number as possible; this will be facilitated by collective tables at the end of the book.

The following procedure is generally adopted for assigning a systematic name to a given compound or for deriving a constitutional/structural formula from a given name:

1. Determine the compound class in question (e. g.: hydrocarbon, heterocycle, carboxylic acid, ketone, halogen derivative, etc., Section 2.2.1, Table 7).
2. Determine the parent structure and define as (possibly further substituted) substituents all other constitutional or nomenclatorial elements present (Chapter 1).
3. As one and the same compound can exhibit the characteristics of more than one compound class and may also be a combination of several parent substructures, seniorities must be laid down, i. e., rules allowing the assignment of priorities.
4. Define which type of nomenclature should be applied, or ascertain which one has been used in a given name. (The IUPAC rules still permit alternative naming possibilities which sometimes overlap with each other, Section 2.2.) It must be noted here that in the domain of substitutive nomenclature – which is always to be preferred – certain traditional class names are no longer considered at all in naming the pertinent individual species, e. g.: ether → alkoxyalkane.
5. The constitutional or nomenclatorial elements separated according to step 3 are individually named (or assigned to partial formulae) and then provided with appropriate locants (numerals, letters) and markers (enclosing marks, hyphens, primes, etc.).
6. Finally, the substituent prefixes, infixes, and suffixes are ordered according to specific rules and then inserted into the name of the parent structure by prefixing them with the appropriate locants.
7. If needed, isotopic modifiers (Section 6.6) and stereochemical descriptors (Section 6.7) must be added.

In connection with the terms "priority" and "seniority" two fundamental principles of systematic nomenclature can be stated:

a) as far as is feasible, lowest locants possible (numerals, letters) should be applied; i. e. when there is a choice, that constitutional or nomenclatorial element is to be preferred which bears the smallest locant.

b) as long as the intended specification remains unambiguous the lowest number of markers (see above) should be used.

Since for many hydrocarbon and heterocyclic parent structures traditional designations are retained, priorities according to the principle of lowest locants and lowest numbers are not always strictly obeyed. In such cases seniorities were assigned more or less arbitrarily.

Since the IUPAC nomenclature system relies totally on the pivotal concept of the "parent structure" to which, in a second sphere, substituents are assigned, it appeared advisable to maintain this division also for the chapters of this book. Thus, we begin with the exposition of the nomenclature rules for parent structures and, in the second chapter, proceed with the discussion of the different types of nomenclature for substituted systems, radicals, and ions; in the third chapter specific classes of functional compounds are addressed, followed, in the forth chapter, by the treatment of metal organyls and, in the fifth, of carbohydrates. The concluding sixth chapter takes up once again the construction of the final names of complex compounds including isotopic modifiers and stereochemical descriptors.

To derive maximum benefit from this book, several technical points should be noted before embarking on its use. To avoid confusion, the term "radical", formerly also used for molecular fragments with pending valences, will now only be employed to designate factual, physical radicals. For nomenclature purposes this term is replaced by the expression "substituent group". Priority criteria are applied successively – when a foregoing criterion is not conclusive, the next one comes into force. The use of hyphens sometimes appears quite arbitrary; if it promotes clarity liberal use thereof might be justified. The examples given are taken from recent research reports or have been devised in such a way that they cover the pertinent rules as completely as possible. If older systematic names are still in use they are considered as "traditional names". The same holds for trivial names in widespread use. When deviations of Chem. Abstr. names from IUPAC names are to be emphasized this will also be duly indicated. In order to familiarize the reader with the extraordinary flexibility of the chemical formula- and sign-language the formula drawings deliberately display the whole spectrum of the graphic possibilities of abstraction.

Literature

The standard IUPAC manual on systematic Nomenclature of Organic Chemistry[1] has recently been supplemented by a "Guide" promulgating current developments in this domain[2]. A wealth of excellent illustrative material concerning the application of systematic nomenclature to ring compounds is presented in the Ring Index[3]. Important topical revisions and extensions of the nomenclature for fused rings[4], von Baeyer polycycles[5], spiro compounds[6], and cyclophanes[7] complement the older compendia. Since in the framework of the registry and index systems of Chemical Abstracts not only unambiguous but to a much greater extent unique names are required, further subtilizations and extensions of the IUPAC rules have been effectuated there. Details are given in the so-called Index Guides attached to the respective five-year indexes[8]. The same holds for the particular interpretations of the IUPAC rules in Beilstein's Handbook of Organic Chemistry – regrettably these are not generally accessible. Detailed rules for the nomenclature of biochemistry and natural products have recently been reissued[9]. Extensive instructions for naming metal- and metalloid-organic compounds can be found in the updated rules of Inorganic Nomenclature[10] as well as in a very recent compilation in Pure and Applied Chemistry[11]. Also very recent is a new, thoroughly revised edition of the nomenclature recommendations for carbohydrates[12]. All new developments and revisions in the area of chemical nomenclature worked out by the IUPAC nomenclature commissions are routinely published in the Journal of Pure and Applied Chemistry.

Closely related to systematic nomenclature are the IUPAC treatises on class names[13] and the terminology of stereochemistry[14], an authoritative article by G. Helmchen dealing exhaustively with all questions relevant to stereochemical notations[15] and a monograph on general chemical terminology[16]. An extensive treatise on status and usage of the language of chemistry has been presented by W. Liebscher[17]. Anyone interested in pertinent historical developments can find abundant information in the books written by W. Holland and edited by V. M. Kisakürek[18] as well as in the collection of articles and documents on the history of organic chemical nomenclature by P. E. Verkade, for many years chairman of the IUPAC nomenclature commission[19]. That the process of name creation and name giving sometimes involves amusing background stories has been impressively demonstrated by Nickon and Siversmith[20].

---

[1] International Union of Pure and Applied Chemistry, Nomenclature of Organic Chemistry, Commission on Nomenclature of Organic Chemistry, Sections A, B, C, D, E, F and H. 1979 Edition. Pergamon Press, Oxford, 1070.

[2] A Guide to IUPAC Nomenclature of Organic Compounds, Recommendations 1993. Blackwell, Oxford, 1993.

[3] A. M. Patterson, L. T. Capell and D. F. Walker: The Ring Index, 2nd. ed. 1960; Supplement I, 1963; II 1964; III 1965. American Chemical Society, Washington, D.C.

[4] IUPAC Recommendations 1998: Nomenclature of Fused and Bridged Fused Ring Systems, (Prepared for publication by G. P. Moss), Pure Appl. Chem. **1998**, *70*, 143.

[5] IUPAC Recommendations 1999: Extension and Revision of the von Baeyer System for naming Polycyclic Compounds (including Bicyclic Compounds), (Prep. for publ. by G. Moss), Pure Appl. Chem. **1999**, *71*, 513.

[6] IUPAC Recommendations 1999: Extension and Revision of the Nomenclature for Spiro Compounds, (Prep. for publ. by G. P. Moss), Pure Appl. Chem. *199*, *70*, 1999.

[7] IUPAC Recommendations 1998: Phane Nomenclature, Part I: Phane Parent Names, (Prep. for publ. by W. H. Powell), Pure Appl. Chem. **1998**, *70*, 1513.

[8] The last thorough changes have been described in section IV of the Index Guide of the Ninth Collective Period (1972–1976). Americal Chemical Society, Chem. Abstr. Service, Columbus, Ohio. In the following Index Guides up to the 13. Coll. Period (1992–1996) further changes have been reported only sporadically.

[9] International Union of Biochemistry and Molecular Biology; Biochemical Nomenclature and related Documents, Portland Press, London,1992. See also: IUPAC Recommendations 1999: Revised Section F; Natural Products and related Compounds, Prep. for Publ. by P. M. Giles, Jr.), Pure Appl. Chem. **1999**, *71*, 587.

[10] IUPAC, Nomenclature of Inorganic Chemisty, Commission on Nomenclature of Inorganic Chemistry. Blackwell, Oxford, 1994.

[11] IUPAC Recommendations 1999: Nomenclature of Organometallic Compounds of the Transition Elements, (Prep. for publ. by A. Salzer), Pure Appl. Chem. **1999**, *71*, 1557.

[12] IUPAC and International Union of Biochemistry and Molecular Biology; Nomenclature of Carbohydrates, (Recommendations 1996, prep. for publ. by A. D. NcNaught), Pure Appl. Chem. **1996**, *68*, 1919.

[13] IUPAC Recommendations 1995: Glossary of Class Names of Organic Compounds and reactive Intermediates, (Prep. for publ. by G. P. Moss, P. A. S. Smith and D. Tavernier), Pure Appl. Chem. **1995**, *67*, 1307.

[14] IUPAC Recommendation 1996: Basic Terminology of Stereochemistry, (Prep. for publ. by. G. P. Moss), Pure Appl. Chem. **1996**, *68*, 2193.

[15] G. Helmchen: Nomenclature and Vocabulary of Organic Stereochemistry in: Houben-Weyl, Methods of Organic Chemistry. Ed. G. Helmchen, R.W. Hofmann, J. Mulzer, E. Schaumann, Stereoselective Synthesis, Vol. E 21 a. Thieme, Stuttgart, 1995, p 1.

[16] M. Orchin, F. Kaplan, R. S. Macomber, R. M. Wilson, H. Zimmer: The Vocabulary of Organic Chemistry. Wiley, New York, 1980.

[17] W. Liebscher: Entwicklung der Fachsprache Chemie. Möglichkeit zur Vereinfachung der Handhabung der Nomenklatur. Habilitationsschrift, Universität Dresden 1991. See also: D. Hellwinkel: Der derzeitige Status der Chemischen Fachsprache, Chemie für Labor und Betrieb **1977**, *28*, 130.

[18] W. Holland: Die Nomenklatur in der Organischen Chemie. Verlag Harri Deutsch, Frankfurt 1969; M. V. Kisakürek (Ed.): Organic Chemistry; its Language and its State of the Art. VHCA, Basel, VCH, Weinheim, 1993.

[19] P. F. Verkade: A History of the Nomenclature of Organic Chemistry. Reidel, Dordrecht, 1985.

[20] A. Nickon, E. F. Silversmith: Organic Chemistry; The Name Game. Modern Coined Terms and Their Origins. Pergamon Press, Oxford, 1987.

1 Parent Structures

1.1
Acyclic Hydrocarbon Systems

1.1.1
Linear Systems

Saturated hydrocarbons are classified by the parent name **alkanes;** substituent groups derived from them are called **alkyl** (or **alkanyl,** see below) groups. The naming system is based on the unbranched members of the homologous series C_nH_{2n+1} of which only the first four are designated by trivial names.

1 Methane, 2 Ethane, 3 Propene, 4 Butane

Starting with $n = 5$ the names are formed systematically by attaching the suffix **...ane** to a numerical term derived from a Greek or Latin numeral.

5 Pentane, 6 Hexane, 7 Heptane, 8 Octane, 9 Nonane

All other unbranched hydrocarbons can be named by combining numerals of the first decade with the respective numerals of the following decades; hundreds and thousands are analogously incorporated into this system.

1 Hen	10 Decane	100 He	1000 Ki
2 Do	20 Cosane	200 Di	2000 Di
3 Tri	30 Tria	300 Tri	3000 Tri
4 Tetra	40 Tetra	400 Tetra	4000 Tetra
5 Penta	50 Penta	500 Penta	5000 Penta
6 Hexa	60 Hexa	600 Hexa	6000 Hexa
7 Hepta	70 Hepta	700 Hepta	7000 Hepta
8 Octa	80 Octa	800 Octa	8000 Octa
9 Nona	90 Nona	900 Nona	9000 Nona

1 Hen ... 9 Nona } { 10 Decane, 20 Cosane, 30 Tria ... 90 Nona } contane { 100 He ... 900 Nona } ctane { 1000 Ki ... 9000 Nona } liane

Exceptions: 1 = mono, 2 = di, 11 = Undecane, 20 = Icosane, 21 = Henicosane. The corresponding substituent groups bear the end-syllable **...yl** instead of ... **ane**; in this case the free valence is always assigned locant 1 (that often can be ommitted). There is a growing tendency, however, to eventually use only the unabbreviated ending ... **anyl**, which could then also be given locants other than 1, e.g.: 2-Methylpropan-2-yl instead of 1,1-Dimethylethyl (= *tert*-Butyl).

$C_{22}H_{46}$ Docosane, $C_{44}H_{90}$ Tetratetracontane,
$-C_{75}H_{151}$ Pentaheptacontyl, $-C_{121}H_{243}$ Henicosahectyl,
$C_{9876}H_{19754}$ Hexaheptacontaoctactanonaliane

Unsaturated acyclic hydrocarbons with double and/or triple bonds are generally designated as alkenes, alkynes, and alkenynes; multiple unsaturation is indicated by the numerical prefixes **di, tri**, etc.: **alkatrienes, alkenediynes** etc.

The class names of the corresponding substituent groups are formed by suffixing the syllable ... **yl: alkadienetriynyl** etc.

To name individual members the principles outlined for saturated systems are applied accordingly; trivial names are retained only for **Methylene** CH_2, **Ethylene** $H_2C=CH_2$, **Allene** $H_2C=C=CH_2$, and **Acetylene** $HC\equiv CH$.

Chains are numbered in such a way that multiple bonds are assigned the smallest numbers possible; if there is a choice double bonds are given the lowest numbers.

$H_3\overset{5}{C}-\overset{4}{C}H_2-\overset{3}{C}H=\overset{2}{C}H-\overset{1}{C}H_3$ Pent-2-ene, $H_3C-C\equiv\overset{2}{C}-CH_3$ But-2-yne

$H_2C=CH-CH_2-C\equiv\overset{2}{C}-\overset{1}{C}H=CH_2$ Hepta-1,6-dien-3-yne

$H_3C-CH=\overset{3}{C}H-C\equiv\overset{1}{C}H$ Pent-3-en-1-yne but:

$H_3\overset{1}{C}-\overset{2}{C}H=CH-\overset{4}{C}\equiv C-CH_3$ Hex-2-en-4-yne

$H_2\overset{1}{C}=\overset{2}{C}=CH-C_{30}H_{61}$ Tritriaconta-1,2-diene

In the corresponding substituent groups the position of attachment is again assigned locant 1, only then are multiple bonds ordered as shown above:

$HC\equiv C-$ Ethynyl
$HC\equiv C-CH=CH-(CH_2)_{20}CH_2-$ Pentacos-22-ene-24-ynyl.

Trivial designations are retained for the groups **Vinyl** $H_2C=CH-$ (systematic name: Ethenyl) and **Allyl** $H_2C=CH-CH_2-$ (systematic name: Prop-2-enyl).

1.1.2
Branched Systems

Branched saturated and unsaturated acyclic hydrocarbons are named as follows: after the main chain has been identified the side chains are attached thereto as substituents. To define the parent chain the following order of seniority is to be observed (see also Section 6.1, p. 178):

a) the main chain must have the maximum number of double and triple bonds together.
b) if the foregoing criterion still leaves of choice or, as in the case of saturated systems, is irrelevant, the highest number of C atoms is decisive.
c) if there is still a choice the highest number of double bonds defines the main chain.
d) if a decision still proves impossible, then that chain which has the largest number of side chains takes precedence.

The main chain thus defined is then numbered according to the principles already specified for unbranched systems. Side chains which in turn can bear side chains of their own are treated analogously. If connecting the partial structures offers a choice, lowest possible locants are given to linking positions. If, in the case of multiple side chains, all the above priority criteria are exhausted, alphabetical order comes into play. This also holds for the citation order of side chains, irrespective of their connectivity locants.

2,3,5-Trimethylhexane

6-Methyl-5-propylundecane

4-Ethyl-5-methylocta-2,6-diene

5-Ethyl-4-methyloct-2-ene-6-yne

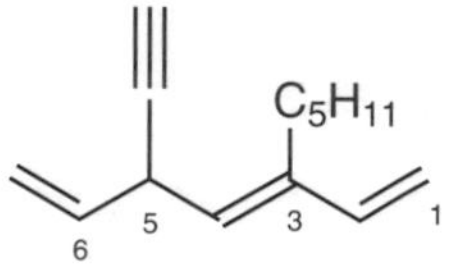

5-Ethynyl-3-pentylhepta-1,3,6-triene

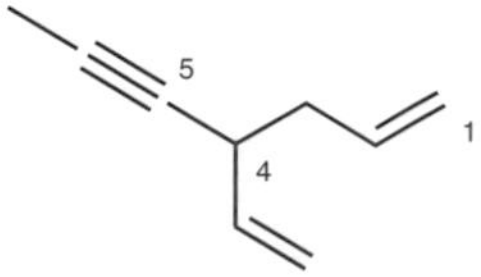

4-Vinylhept-1-ene-5-yne

For simply branched hydrocarbons and their substituent groups the following trivial designations are retained

Isobut $\begin{cases} \text{ane} \\ \text{yl} \end{cases}$, Isopent $\begin{cases} \text{ane} \\ \text{yl} \end{cases}$, Isohex $\begin{cases} \text{ane} \\ \text{yl} \end{cases}$ $(CH_3)_2CH-(CH_2)_n-CH_3$

respectively

$(CH_3)_2CH-(CH_2)_n-CH_2-$ $n = 0-2$

Neopent $\begin{cases} \text{ane} \\ \text{yl} \end{cases}$ $(CH_3)_4C$ resp. $(CH_3)_3C-CH_2-$

Isoprene

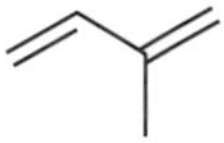

Isopropyl $(CH_3)_2CH-$,   Isopropenyl

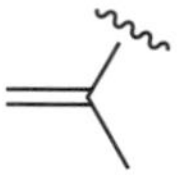

sec-Butyl 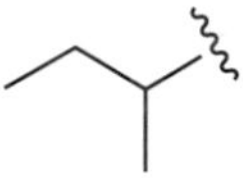   tert-Butyl $(CH_3)_3C-$

Instead of the alphanumerical group designations $-CH_3$, $-C_2H_5$, $-C_3H_7$, $-C_4H_9$ the alphabetical short forms $-$**Me, **$-$**Et, **$-$**Pr, **$-$***i*Pr, **$-$**Bu, **$-$***i*Bu, **$-$***s*Bu, **$-$***t*Bu** can be employed analogously.

1.1.3
Systems with Branched Side Chains

If further branched side chains are present it must be remembered that side chains are always connected through their position 1 with the main chain (see also note on p. 8) and that their alphabetical order is determined by the first letter of the name of the complete substituted substituent. If a choice then still remains, lower numbers within the substituents will become decisive.

$$H_3C-\overset{1}{C}H-\overset{2}{C}H-CH_2-CH_2-CH_3$$
$$\overset{13}{H_3C}-(CH_2)_5-\underset{7}{C}H-CH_2-\overset{5}{C}H-CH_2-CH_2-CH_2-\overset{1}{C}H_3$$

with CH₃ groups on positions 2 and the lower chain CH₃–CH–CH₃

7-(1,2-Dimethylpentyl)-5-(isopropyl)tridecane

$$H_3C-CH_2-\overset{2}{C}H-\overset{1}{C}H_2 \qquad \overset{1}{C}H-CH_2-CH_2-CH_3$$
$$\overset{13}{H_3C}-(CH_2)_4-\overset{8}{C}H-CH_2-\overset{6}{C}H-(CH_2)_4-\overset{1}{C}H_3$$

with CH₃ substituents

6-(1-Methylbutyl)-8-(2-methylbutyl)tridecane

Identical side chains are indicated by the multiplicative prefixes **di, tri, tetra, penta,** etc. If side chains with identical further substituents are present the multiplicative prefix forms **bis, tris, tetrakis, pentakis,** etc. are employed. (The prefix forms **bi, ter, quater, quinque, sexi, septi,** etc. are reserved for direct linking of identical units; see the following Sections.)

$$H_3C-CH_2-CH_2-\overset{1}{C}-CH_3$$
$$\overset{10}{H_3C}-CH_2-CH_2-CH_2-CH_2-\overset{5}{C}-CH_2-CH_2-\overset{2}{C}H-\overset{1}{C}H_3$$
$$H_3C-CH_2-CH_2-\overset{1}{C}-CH_3$$

with CH₃ substituents

5,5-Bis(1,1-dimethylbutyl)-2-methyldecane

If such a complex hydrocarbon system is present as a substituent group, the free valence must again be assigned locant 1; only then are the usual prority rules applied accordingly.

$$H_3C-CH=\overset{2}{C}-\overset{1}{C}H_2-\{$$
$$\overset{1}{C}H_2-CH_2-\overset{3}{C}H-(CH_2)_5-CH_3$$
$$C_5H_{11}$$

2-(3-Pentylnonyl)but-2-enyl

1.1.4
Multivalent Substituent Groups

Multivalent substituent groups of acyclic hydrocarbons are designated by
attaching the suffixes **...idene** and **...idyne** to the name of the corresponding monovalent group insofar as the free valences are at the same C
atom. Multiple occurrence of such structural elements is taken into
account with suitable multiplicative infixes. **Methylene,** $=CH_2$ or $-CH_2-$, is
retained as trivial name; the group $=CH-$ is sometimes still called **methine**
or **methyne.**

HC$\equiv$ Methylidyne, $H_2C=C=$ Vinylidene, $(CH_3)_2C=$ Isopropylidene,

$\equiv$C–CH$_2$–CH$=$CH–C$\equiv$ Pent-2-ene-1,4-bis(ylidyne),

$\equiv$C–C$\equiv$C–CH$=$ But-2-yne-1-yliden-4-ylidyne,

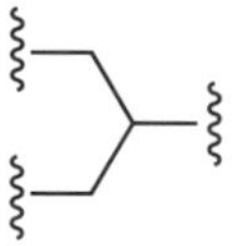

Propane-1,2,3-triyl,

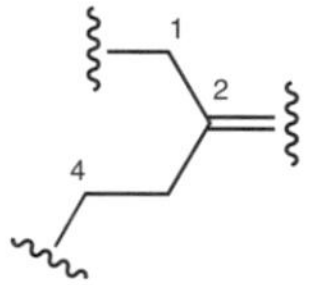

Butane-1,4-diyl-2-ylidene,

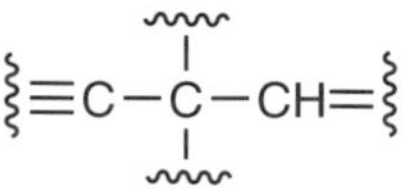

$\equiv$C–C–CH$=$ Propane-2,2-diyl-1-yliden-3-ylidyne.

Hydrocarbon chains with free valences at each terminal C atom are frequently still called **trimethylene-, tetramethylene-,** etc. instead of **alkane-
1-ω-diyl** groups, the systematically correct designation. Unsaturated substituent groups of this kind are named by replacing the terminal syllable
...ene with **...enylene.** The trivial terms **ethylene** for $-CH_2-CH_2-$, **propylene** for $H_3C-CH-CH_2-$, and **vinylene** for $-CH=CH-$ are still frequently

encountered. For more complex derivatives thereof, systematic names should always be given preference.

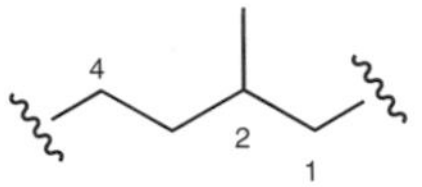

2-Methylbutane-1,4-diyl
(2-Methyltetramethylene)

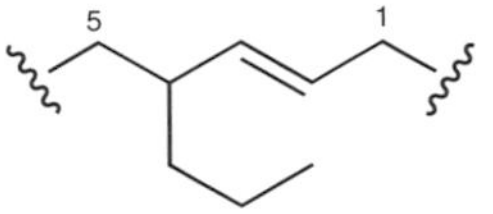

4-Propylpent-2-ene-1,5-diyl
(4-Propylpent-2-enylene)

1.2
Cyclic Systems

1.2.1
Cyclic Hydrocarbon Systems

1.2.1.1
Monocyclic Hydrocarbons

Saturated and unsaturated monocyclic hydrocarbons are treated like their acyclic analogues but with the specifying prefix **cyclo** in front of the name. Monovalent substituent groups derived therefrom are again given the ending **...yl,** bivalent groups the ending **...diyl** (formerly: **...ylene**) when different C atoms are involved and **...ylidene** or **...1,1-diyl** when the free valences are at the same carbon atom. In numbering them, positions with free valences have precedence; only then are double and triple bonds together and thereafter double bonds given the lowest locants possible.

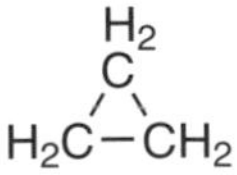

Cyclopropane

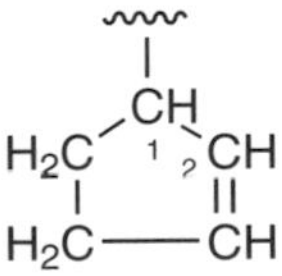

Cyclopent-2-enyl

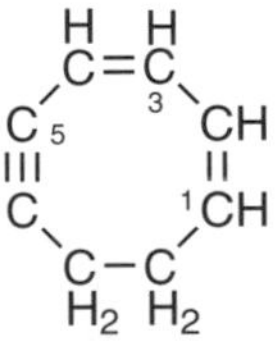

Cycloocta-1,3-dien-5-yne

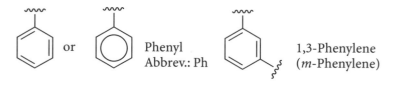

Cyclobutane-1,3-diyl   Cyclohex-4-ene-1,2-diyl   Cyclohept-6-en-
formerly: formerly: 2-yne-1-yl-4-ylidene
1,3-Cyclobutylene   Cyclohex-4-en-1,2-ylene

The designations **benzene, phenyl** and ***o,m,p*-phenylene** are retained as
trivial names.

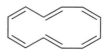   or   Phenyl   1,3-Phenylene
                                   Abbrev.: Ph   (*m*-Phenylene)

Monocyclic polyenes containing the maximum number of non-cumulative
double bonds (for short called: **mancude** systems) and having the general
composition C_nH_n or C_nH_{n+1} ($n > 6$) can also be named as **[*n*]annulenes.**

[12]Annulene   1*H*-[9]Annulene
(Cyclododeca-   (Cyclonona-1,3,5,7-tetraene)
1,3,5,7,9,11-hexaene)

1.2.1.2
Polycyclic Hydrocarbons

1.2.1.2.1
Fused Polycyclic Hydrocarbons

The nomenclature of these compounds in which at least two highly unsaturated rings are fused together through at least two common C-atoms
is based on an extended series of trivial names. The most important of

these are listed in Table 1 in ascending order of seniority. Systems comprising only benzene units and their substitution products are generally designated as **arenes** or, traditionally, **aromatics**. Fused hydrocarbon systems for which no trivial names are retained are systematically named as follows:

That component that has the largest ring or/and that – if it is a trivial system according to Table 1 – contains the largest number of rings is defined as **parent** (or base or primary or main) component. All other components are attached in the form of prefixes to the parent name by changing their ending ... **ene** to ... **eno**. The following abbreviated prefix forms are retained:

Acenaphtho	from Acenaphthylene	Naphtho	from Naphthalene
Anthra	from Anthracene	Perylo	from Perylene
Benzo	from Benzene	Phenanthro	from Phenanthrene

With the exception of **benzo,** fusion prefixes for monocyclic systems are treated as exemplified for **cyclopenta, cyclohepta** etc. It should be noted here that for fused polycyclic systems the ending ... **ene** always indicates the **maximum number of non-cumulative double bonds,** that is, a **mancude system!**

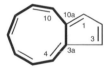

Cyclopentacyclononene
or Cyclopenta[9]annulene
not: Cyclopentadieno-
cyclonononatetraene

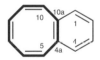

Benzocyclooctene
or Benzo[8]annulene
not: Benzocyclooctatetraene

For the following compounds Chem. Abstr. incomprehensibly uses the von Baeyer names **bicyclo[4.2.0]octa-1,3,5,7-tetraene** and **1H-bicyclo[4.1.0] hepta-1,3,5-triene** that totally disregard the aromatic character of these compounds.

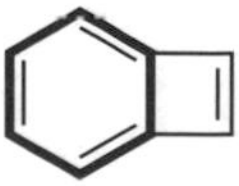

Cyclobutabenzene 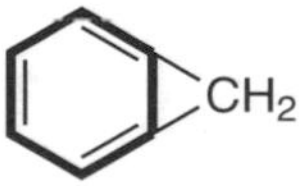   Cyclopropabenzene

Table 1. Retained trivial names for fused polycycles in ascending priority order (see also Section 6.3)

1 Pentalene	7 *as*-Indacene
2 Indene	8 *s*-Indacene
3 Naphthalene	9 Acenaphthylene
4 Azulene	10 Fluorene
5 Heptalene Analogously: Octalene etc.	11 Phenalene
6 Biphenylene	12 Phenanthrene[a]

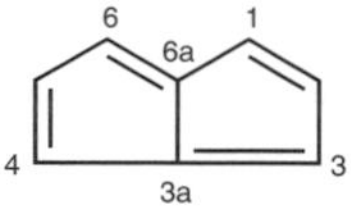

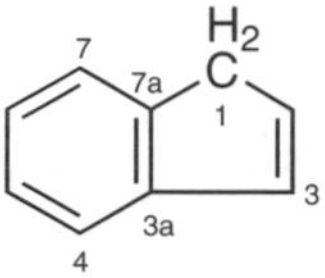

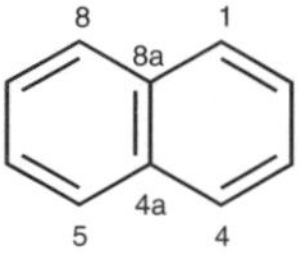

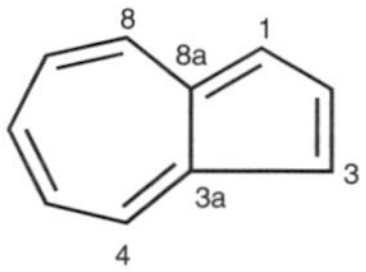

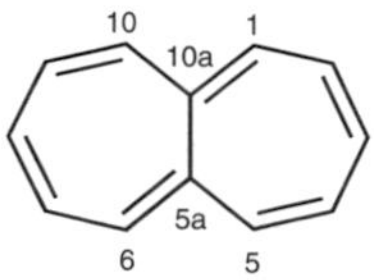

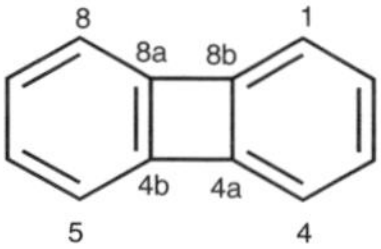

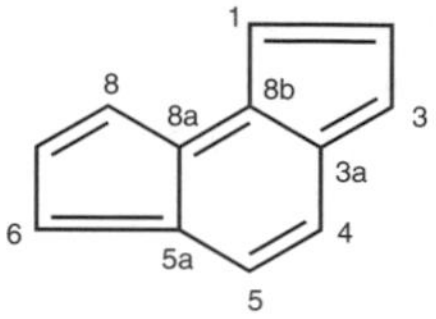

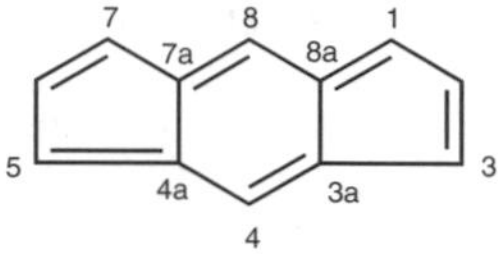

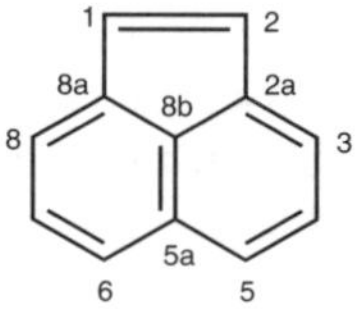

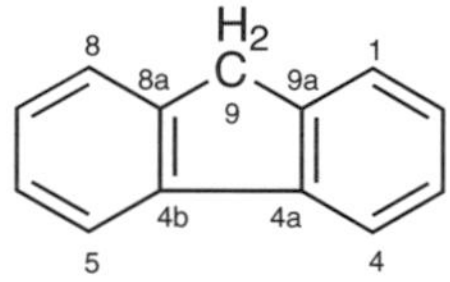

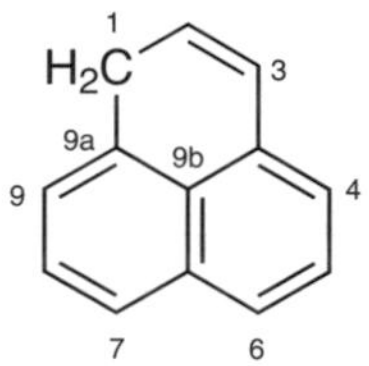

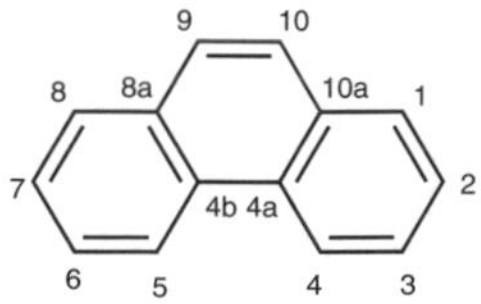

Table 1 (continued)

13 Anthracene[a]

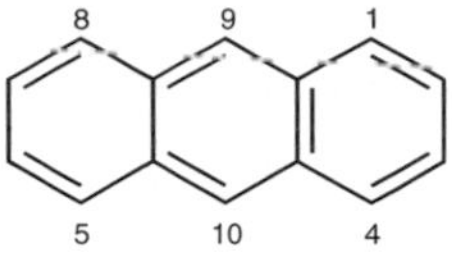

14 Fluoranthene

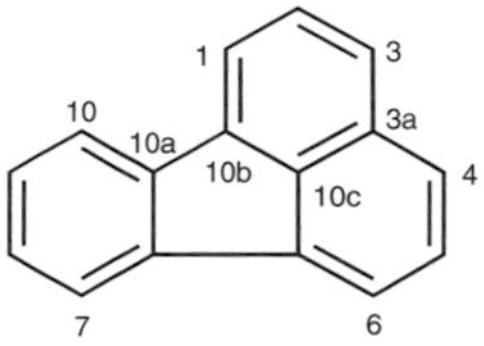

15 Acephenanthrylene

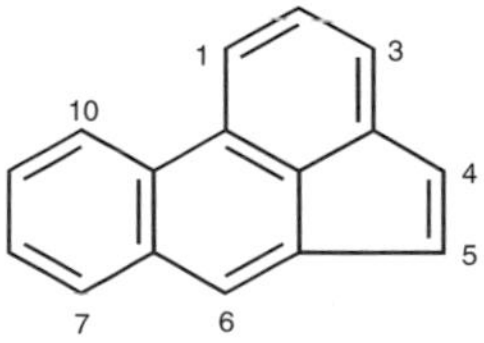

16 Aceanthrylene

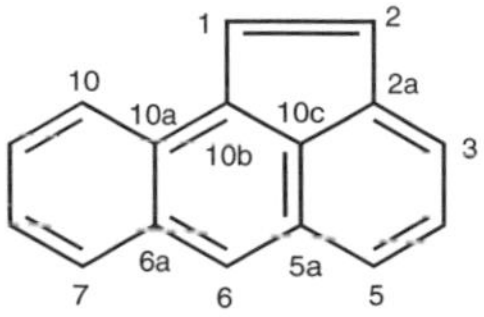

17 Triphenylene[b]

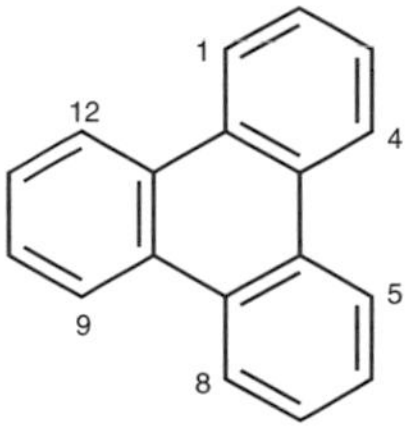

18 Pyrene

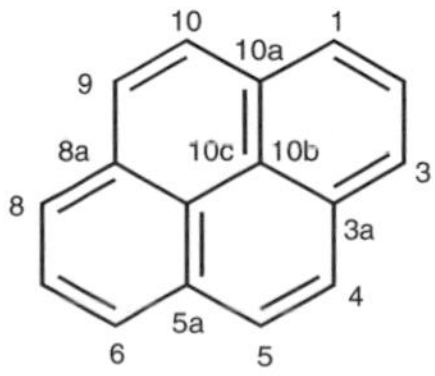

Table 1 (continued)

19 Chrysene	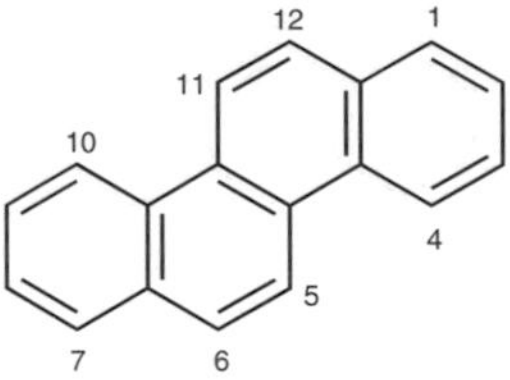
20 Naphthacene Now: Tetracene	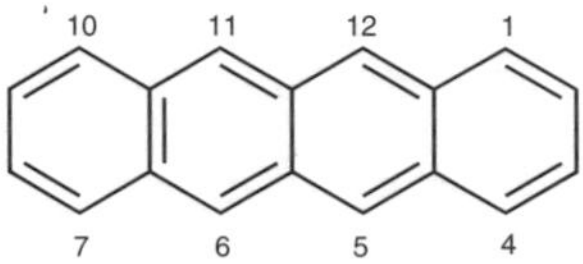
21 Pleiadene	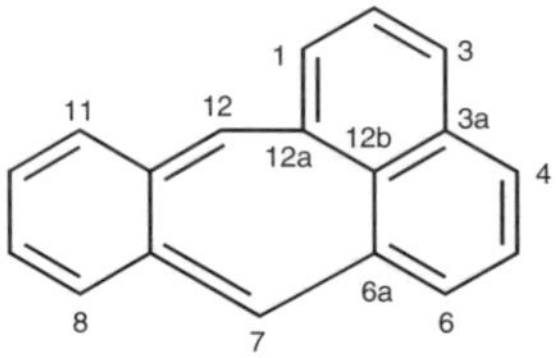
22 Picene	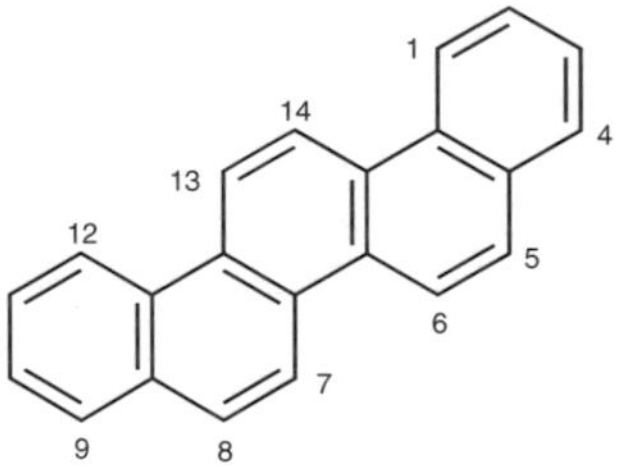
23 Perylene	

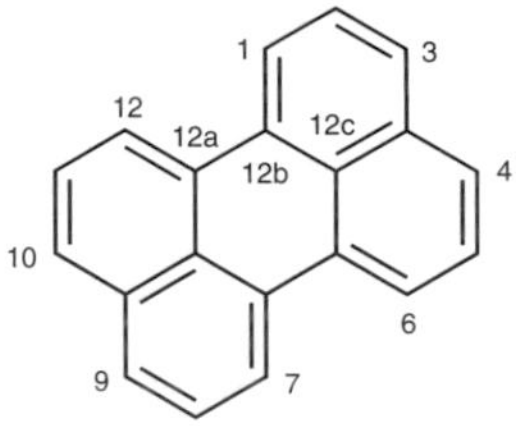

Table 1 (continued)

24 Pentaphene	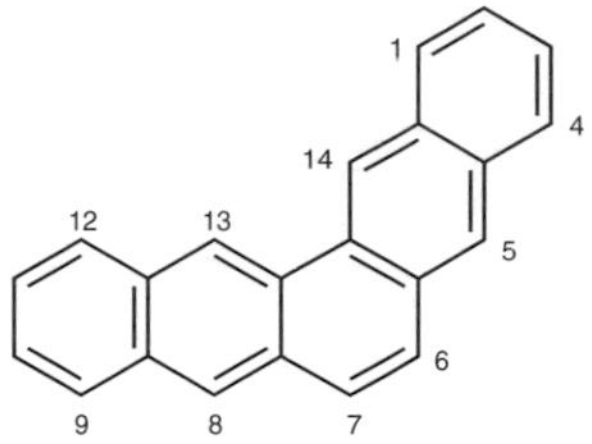
25 Pentacene	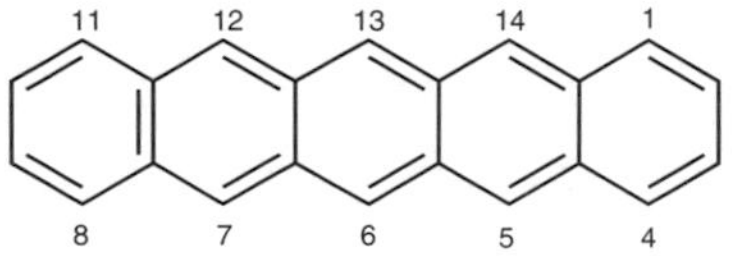
26 Tetraphenylene[b]	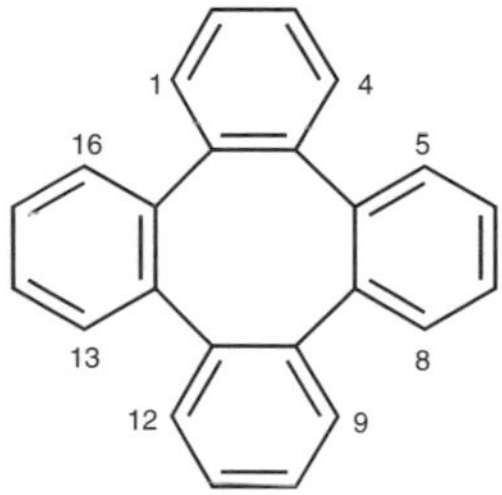
27 Hexaphene	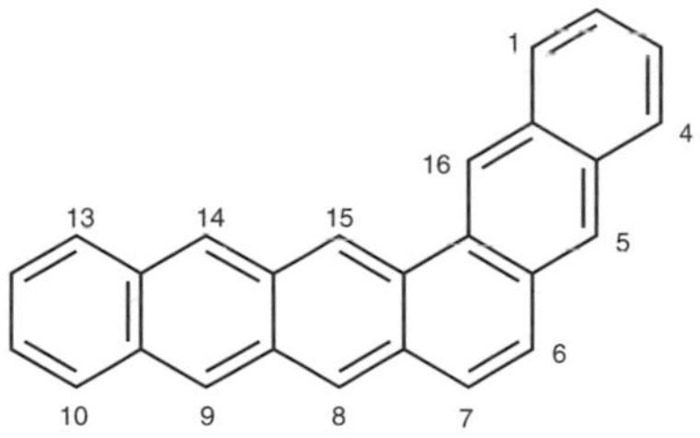
28 Hexacene	

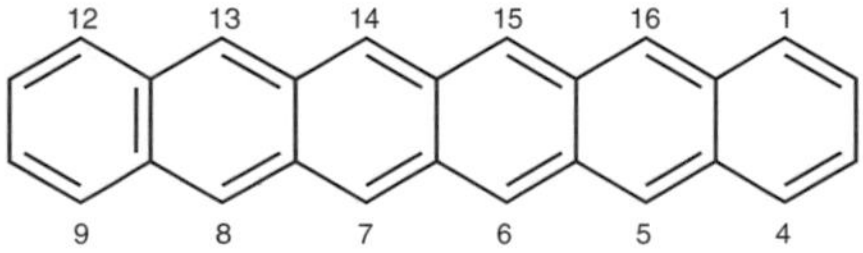

Table 1 (continued)

29 Rubicene

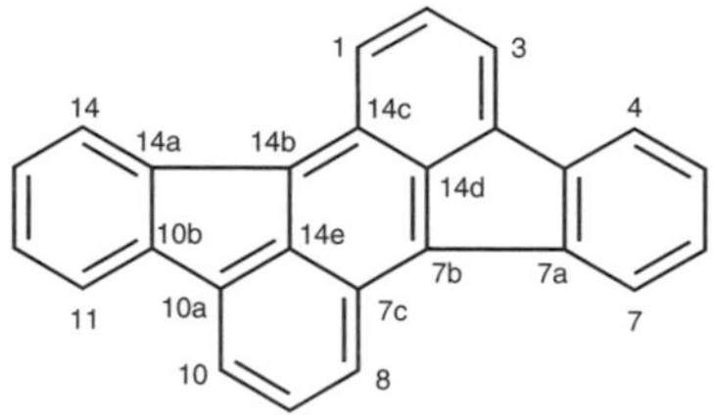

30 Coronene

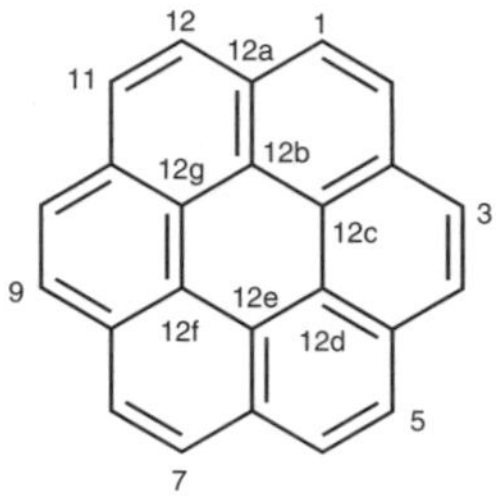

31 Trinaphthylene

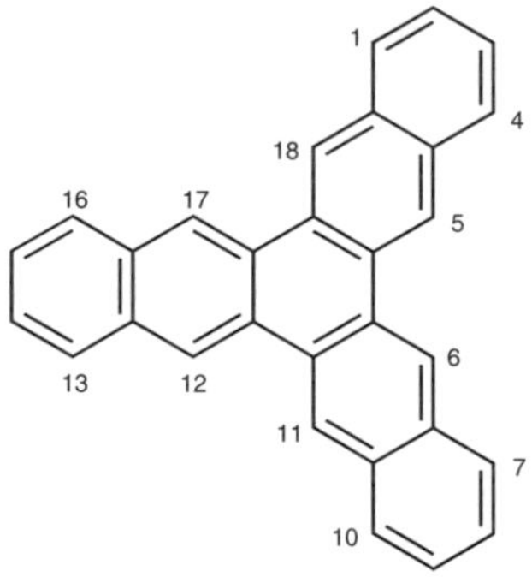

32 Heptaphene

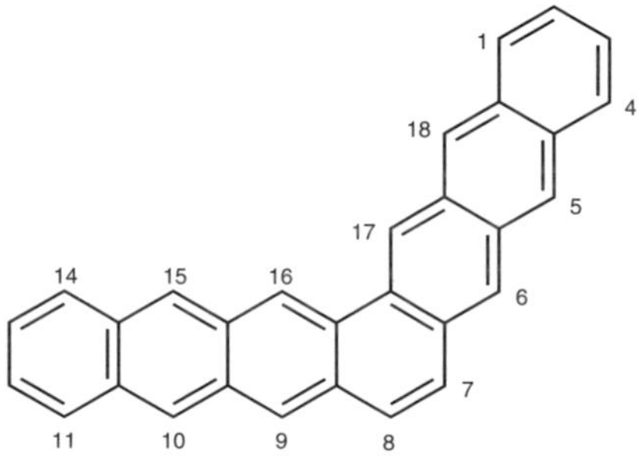

Table 1 (continued)

33 Heptacene[c]	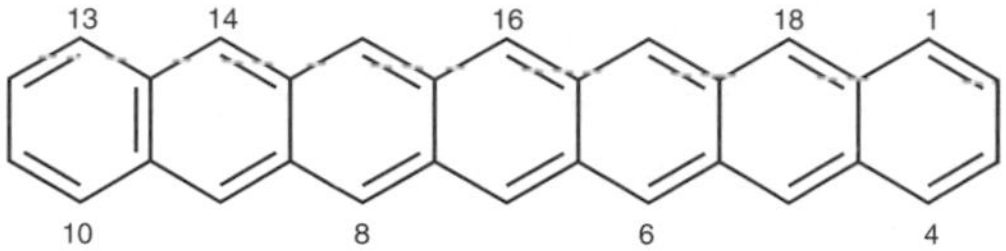
34 Pyranthrene	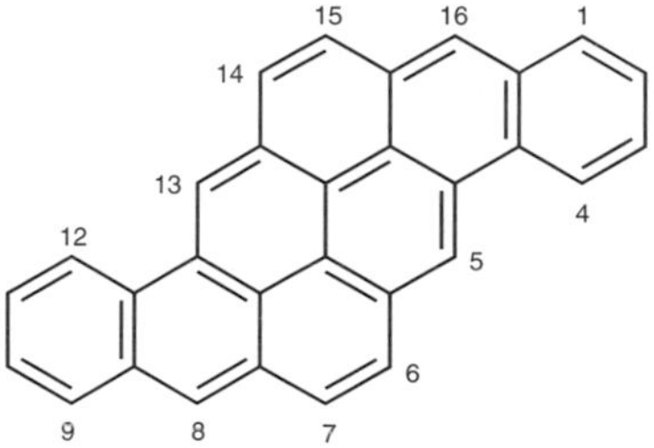
35 Ovalene	

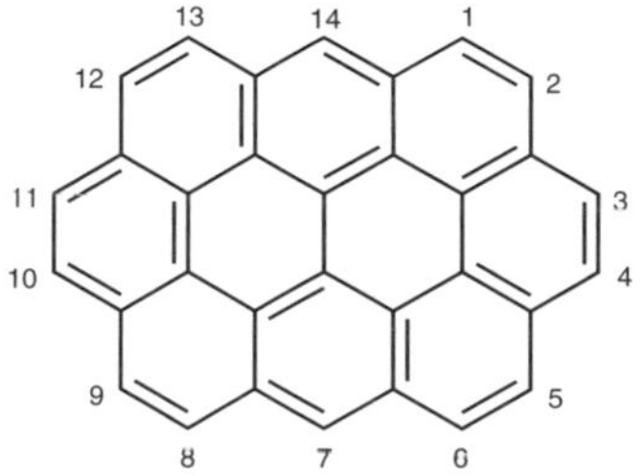

[a] Exceptions from systematic numbering.

[b] Should really be terphenylene, quaterphenylene etc. because the direct joining of identical components (here o-phenylene) is normally indicated by **bi, ter, quater** etc. In addition, this naming procedue is also applicable for higher and *meta*-linked representatives of this series: pentaphenylene, hexa-*m*-phenylene, etc.

[c] Can be continued accordingly: octacene, nonacene... polyacene; likewise: ...polyaphene.

Generally, many isomers are possible for fused systems and these have to be differentiated by appropriate descriptors. To obtain these, the sides of the parent component are labelled consecutively by italic letters *a*, *b*, *c*... tracing the locant path 1, 2, 3, 4 ... and ignoring potential non-standard numbering. For the attached secondary components the inherent numbering is maintained. Combination of the **partial names** is then effected by intercalating the fusion locants in square brackets between the component names. Clearly, the principle of lowest locants is to be respected here too: first, lowest letters for the base component; second, lowest numbers for the higher order components; the sequence of numbers follows from the sequence of letters! If several attached (secondary) components are present alphabetical order is valid as usual.

In order to find the correct sequences of letters and numbers for both the individual components and the final assembly of a fused system it is imperative to follow a series of orientation rules as outlined in the following.

First of all, a uniform graphical representation of all the rings involved in the construction of a fused assembly must be decided upon as shown below:

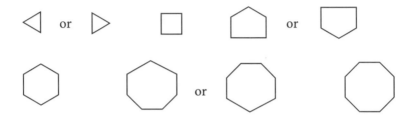

Vertically drawn bonds must be distanced as far as possible from each other; this is of particular significance for higher membered rings, e.g.:

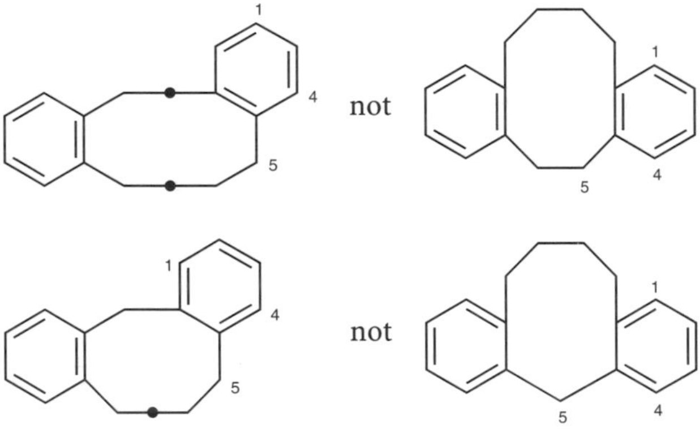

The formulae are subsequently oriented in such a way that:

a) as many rings as possible are positioned on a horizontal axis,
b) as many rings as possible occur in the upper right quadrant,
c) as few rings as possible are placed in the lower left quadrant,
d) as many rings as possible are situated above the horizontal axis,
e) the priority of numbering is guaranteed.

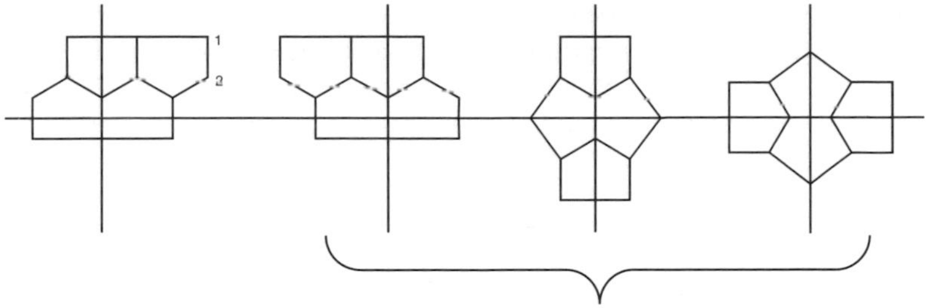

correct   incorrect

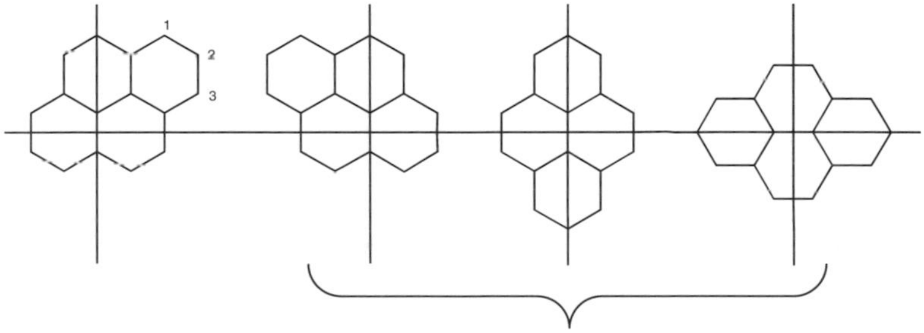

correct   incorrect

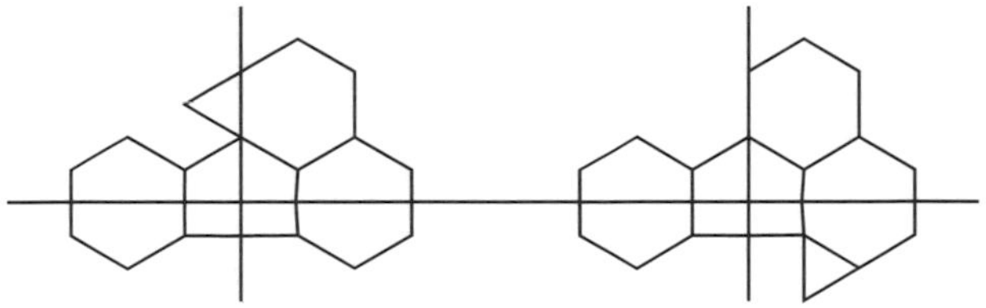

correct   incorrect

The **numbering** starts at the **most counterclockwise** non-fused C atom of the **uppermost ring positioned farthest to the right** and proceeds **clockwise** around the periphery of the whole structure. Carbon atoms common to more than one ring are numbered indirectly by attaching the letters a, b, c, etc. to the locant of the preceding C atom. Until now interior C atoms were numbered in the same manner starting from the highest locant of the periphery and proceeding clockwise towards the center of the structure.

According to a recent rule revision the numbering of each interior atom is now related to its **lowest numbered** peripheral neighbour atom by a superscript denoting the number of bonds between the two atoms. The following example stands among others for the intricacies involved in the application of orientation rules a) through e). Here the decision for the correct orientation is only possible after lowest locants for the fusion positions have been evaluated.

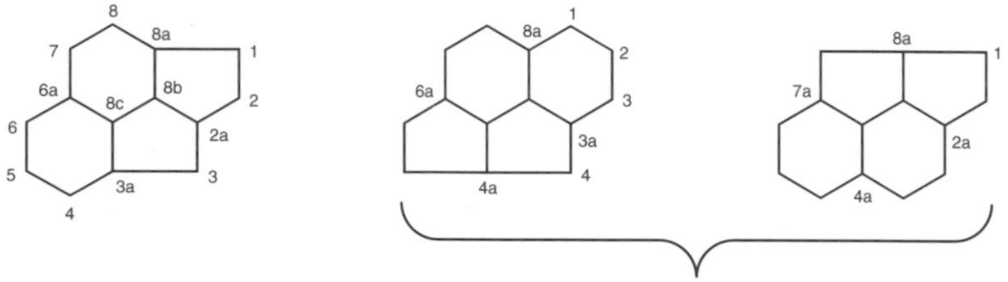

correct   incorrect

The sequence 2, 3, 6, 8 is lower than and thus senior to 3, 4, 6, 8 or 2, 4, 7, 8. Note: not the sum is decisive but the first point of difference.

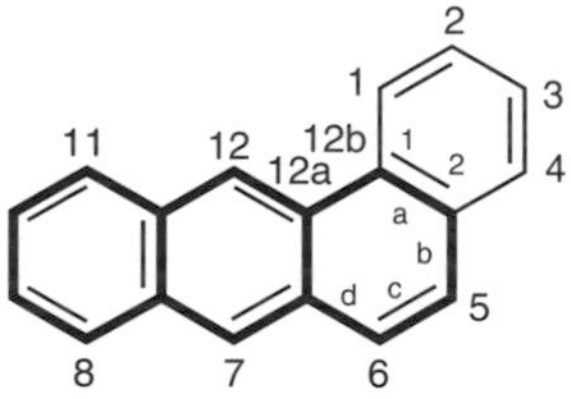

Benzo[*a*]anthracene 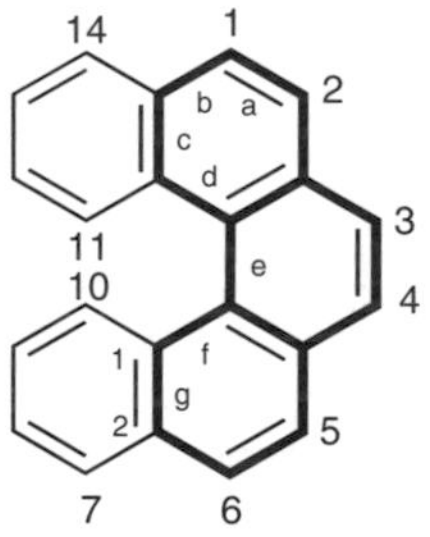   Dibenzo[*c,g*]phenanthrene
                               not: Naphtho[2,1-*c*]phenanthrene

If additional benzo units are likewise *ortho*-fused to the above pentacyclic system, helical structures result which can be specified as **hexa-, hepta-** etc. **-helicenes.** Since the current rules lead to confusing variable peripheral numberings for individual members of that compound class, a new consistent numbering scheme has been proposed according to which numbering always starts at one end of the helix and proceeds sequentially to the other.

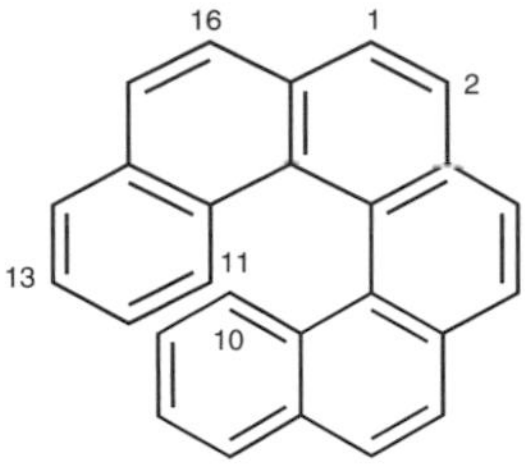

Hexahelicene

New numbering

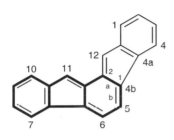

Indeno[2,1-a]fluorene

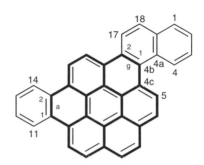

Benzo[a]naphtho[2,1-g]coronene

As in the following two examples the sequence of letters is pre-established in the clockwise sense, the sequence of numbers, in order to guarantee lowest numbers, must run counter-clockwise from the higher to the lower number. Therefore the descriptors must read ... [2,1-a] ... and ... [2,1-g] ... and not ... [1,2-a and g] ..., respectively.

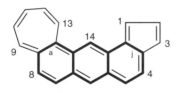

Cyclohepta[a]cyclopenta[j]anthracene

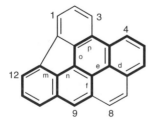

Benzo[def]indeno[1,2,3,4-mnop]chrysene

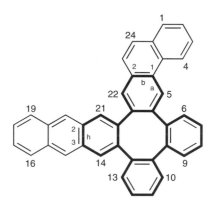

Dinaphtho[1,2-*b*:2,3-*h*]tetraphenylene

In more complex systems containing higher order components attached to the secondary component these are identified with primed numbers within the usual square brackets.

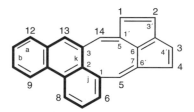

Pentaleno[1′,6′:5,6,7]cycloocta[1,2,3-*jk*]phenanthrene

Naphtho[1″,2″:3′,4′]cyclobuta[1′,2′:3,4]cyclobuta[1,2-*e*]biphenylene

Fused systems where a subordinate central component is fused to two or more identical senior outer components are named as follows:

Cyclobuta[1,2-*a*:3,4-*a*′]diindene

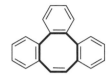

Benzo[1,2:3,4:5,6]triscyclooctene
or: Benzo[1,2:3,4:5,6]tri[8]annulene

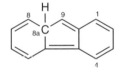

Tribenzo[a,c,e]cyclooctene or
Tribenzo[a,c,e][8]annulene

If a fused hydrocarbon contains saturated positons – ⩾CH and/or >CH$_2$ groups exhibiting **indicated hydrogens** – these are specified with a locant and the italicised letter *H*. When such positions occur pairwise conjugated, the compounds in question are named as **di-, tetra-,** or **perhydroderivatives.** If there is a choice, hydrogenated positions must be given lowest possible locants, indicated hydrogen having precedence and being placed directly in front of the hydrocarbon name.

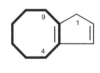

8a*H*-Fluorene   4*H*-Indene   4,5,6,7,8,9-Hexahydro-
                                                  1*H*-cyclopenta-
                                                  cyclooctene

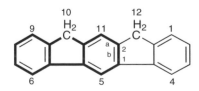

3,4-Dihydrophenanthrene   Perhydropentalene
(exceptional numbering)

10,12-Dihydroindeno[2,1-*b*]fluorene

For a couple of partially saturated fused hydrocarbons the following long established trivial names are retained:

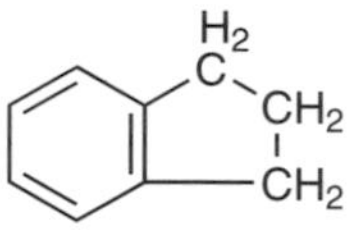

Indane
(2,3-Dihydro-
1H-indene)

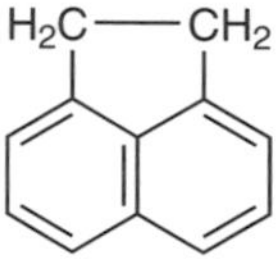

Acenaphthene
(1,2-Dihydro-
acenaphthylene)

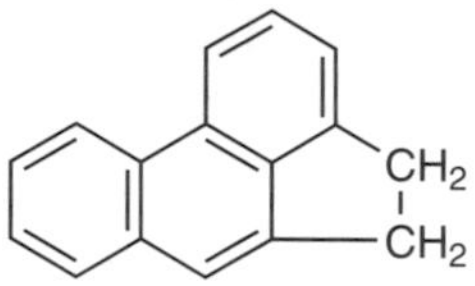

Acephenanthrene
(4,5-Dihydro-
acephenanthrylene)

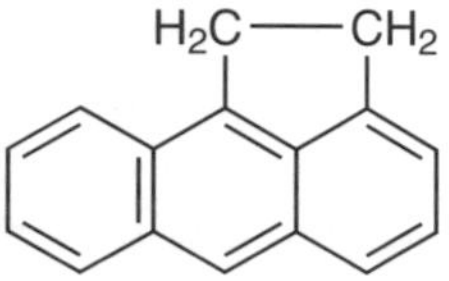

Aceanthrene
(1,2-Dihydroaceanthrylene)

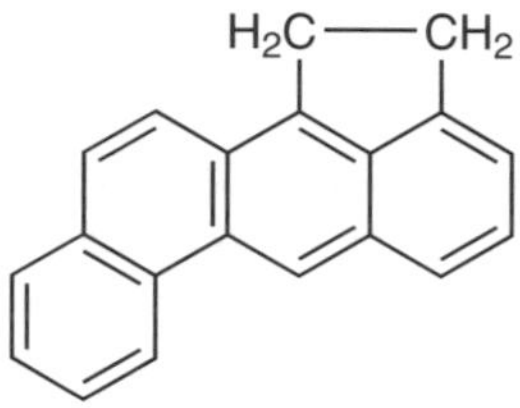

Cholanthrene

The names of substituent groups derived from fused hydrocarbons are obtained, as repeatedly described, by attaching the syllables

yl   for monovalent groups
ylidene   if two free valences are at the same C atom
diyl   if two free valences occur at different C atoms (formerly: **ylene**)

More than two free valences are indicated by **...triyl, ...tetrayl,** etc. If, as in Chem. Abstr. practice, unabbreviated systematic group terms are used instead of the traditional short forms **2-naphthyl, 9-anthryl, 4-phenan-thryl,** etc. the corresponding locants are inserted in front of the function-alities: **naphthalen-2-yl, phenanthren-4-yl,** etc. Within the frame of hydro-carbon numbering, free valences are given the lowest locants possible.

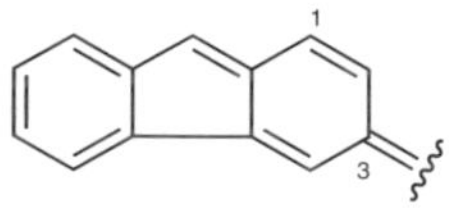

Fluoren-3-ylidene

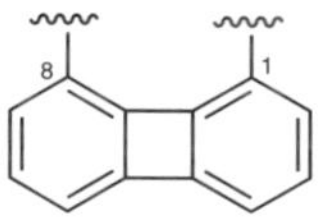

Biphenylene-1,8-diyl
(1,8-Biphenylenylene)

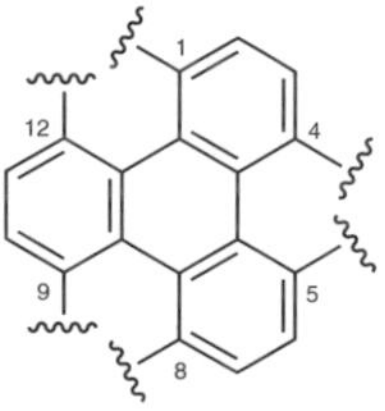

Triphenylene-
1,4,5,8,9,12-hexayl

1.2.1.2.2
Bridged Polycyclic Hydrocarbons

1.2.1.2.2.1
von Baeyer System

Saturated cyclic hydrocarbons containing two or more rings in which at least two rings have at least two common C-atoms are specified as **bi-, tri-, tetra-,** etc. **-cycoalkanes.** The total number of rings present evidently equals the number of C-C cleavages necessary to obtain an open chain compound.

The names of such compounds are constructed stepwise in the following manner:

a) the actual tridimensional structure is transformed into an appropriate planar representation,

b) that ring is selected as **main ring** which includes as many skeletal atoms as possible,

c) the largest possible carbon chain that connects two C-atoms of the main ring (consisting itself of two branches) is defined as the **main bridge.** The corresponding points of connection are called **bridgeheads,**

d) to assemble the full name the numbers of C atoms of **both branches** of the **main ring,** the **main bridge** and the **secondary bridges,** if present, are cited in descending order, separated by full stops, and placed in square brackets after the term **cyclo.**

e) the numbering begins at the first bridgehead, tracing first the largest branch of the main ring up to the second bridgehead and then the next largest branch back to the first bridgehead. Next comes the (longest possible) main bridge which should divide the main ring as symmetrically as possible and then the secondary bridges. The bridgeheads of the secondary bridges are specified by superscripts chosen as small as possible in the framework of the above rules.

Bicyclo[4.4.0]decane
(Decalin)

Bicyclo[3.3.1]nonane

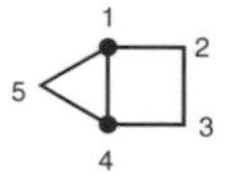

Bicyclo[2.1.0]pentane

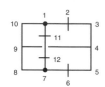 ≡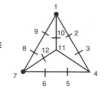

Tricyclo[5.3.2.0^{4,9}]dodecane
correct

...[5.2.3.0^{4,11}]...
incorrect

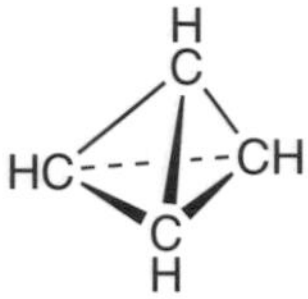

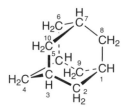 ≡ ≡

Tetrahedrane 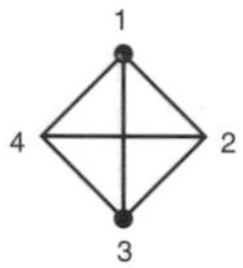 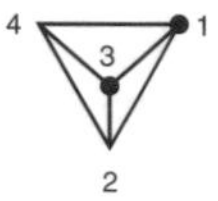   Tricyclo[1.1.0.0^{2,4}]butane

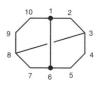 ≡ ≡

Adamantane 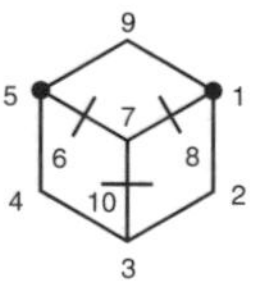 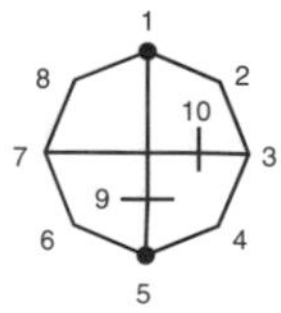   Tricyclo[3.3.1.1^{3,7}]decane

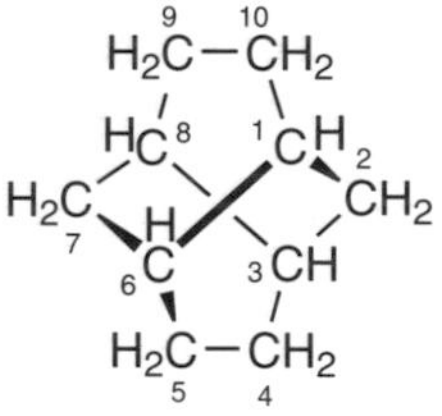

 ≡

Twistane   Tricyclo[4.4.0.0^{3,8}]decane

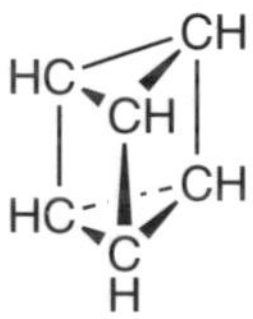

Prismane 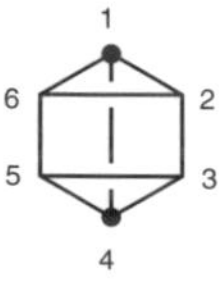 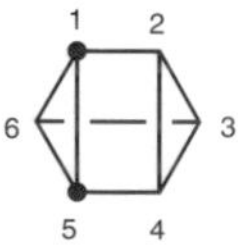   Tetracyclo[2.2.0.0^{2,6}.0^{3,5}]-   Tetracyclo[3.1.0.0^{2,4}.0^{3,6}]-
              hexane (correct)   hexane (incorrect)

Symmetrical partition of the main ring

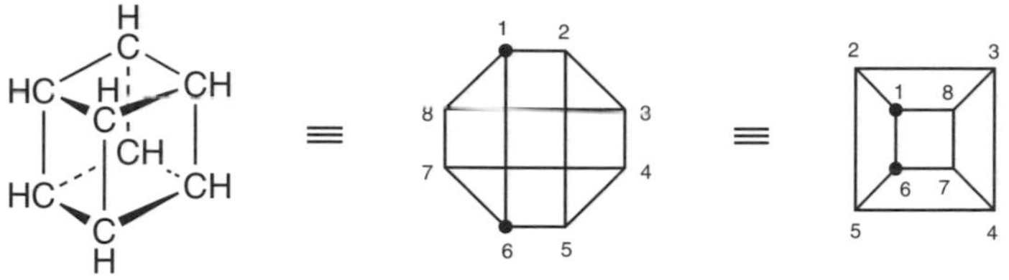

Cubane: Pentacyclo[4.2.0.0^{2,5}.0^{3,8}.0^{4,7}]octane

On insertion of double bonds, especially more than one such bond, into such systems one rapidly encounters borderline situations which in many cases suggest that it might be more reasonable to apply an alternative naming method. While the following compounds, both valence isomers of benzene (as is prismane above), can be named clearly and unambiguously by the von Baeyer system,

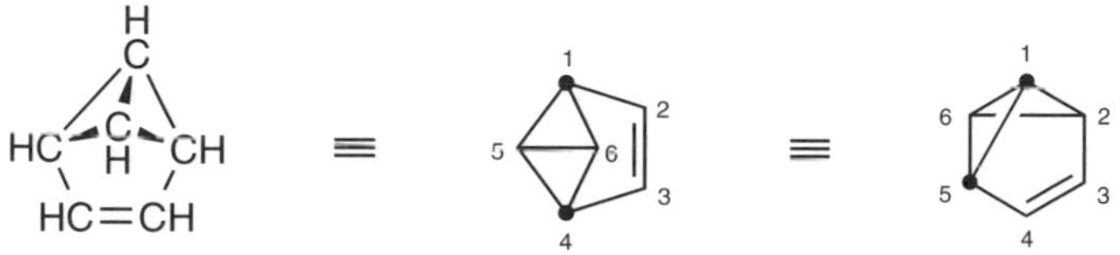

Benzvalene

Tricyclo[2.1.1.0^{5,6}]
hex-2-ene
incorrect

Tricyclo[3.1.0.0^{2,6}]
hex-3-ene
correct

Dewar benzene

Bicyclo[2.2.0]hexa-2,5-diene

for compounds containing intact benzenoid skeletons the nomenclature of fused polycycles or of parent structures substituted by side chains is much more reliable and recommendable.

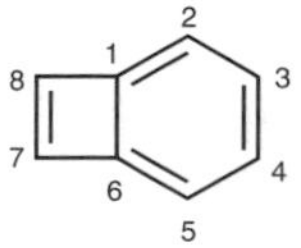

Bicyclo[4.2.0]octa-1,3,5,7-
tetraene

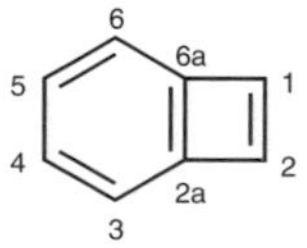

better: Cyclobutabenzene

$$\overset{16}{HC}=\overset{1}{C}-(CH_2)_5-\overset{7}{C}H_2$$
$$\overset{|}{CH_{17}}$$
$$\overset{||}{CH_{18}}$$
$$\overset{|}{HC}=\overset{|}{C}-(CH_2)_5-CH_2$$
$$\overset{}{15}\quad\overset{}{14}\qquad\qquad\qquad\overset{}{8}$$

$$\equiv$$

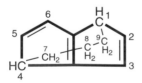

$(CH_2)_{12}$

Bicyclo[12.2.2]octadeca-1(16),14,17-triene

better: 1,4-Dodecamethylenebenzene
or 1,4-Dodecanobenzene (see below)
or (1,4)Benzenacyclotridecaphane
(**Cyclophane nomenclature** p. 68)

1.2.1.2.2.2
Bridged Fused Systems

For more or less extended fused systems with additional bridges the von Baeyer system has been totally abandoned. Here the bridge designators derived by converting the hydrocarbon names ...**ane**, ...**ene** to ...**ano**, ...**eno** are **attached** as prefixes and in alphabetical order to the name of the parent structure. The corresponding bridgeheads are associated in the form of lowest possible locants and the bridge members enumerated consecutively from the bridgehead bearing the highest locant. The most conspicuous criteria for selecting the underlying parent structure of a bridged fused system are examined one after another as follows:

a) maximum number of rings,
b) maximum number of skeletal atoms,
c) lowest number of hetero atoms (see also p. 67),
d) most senior ring system.

1,4-Dihydro-1,4-propanopentalene

Perhydro-1,4-ethano-5,8-methanoanthracene

The trivalent bridge HC⋜ is (still) named **metheno-** although systematically it would be called **methanetriyl.** For unsaturated bridges, locants of multiple bonds are quoted in square brackets within the bridge name.

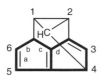

1,2,4,6a-Tetrahydro-1,2,4-methe-
nocyclopenta[cd]pentalene
better:
... 1,2,4-(methanetriyl)cyclo ...

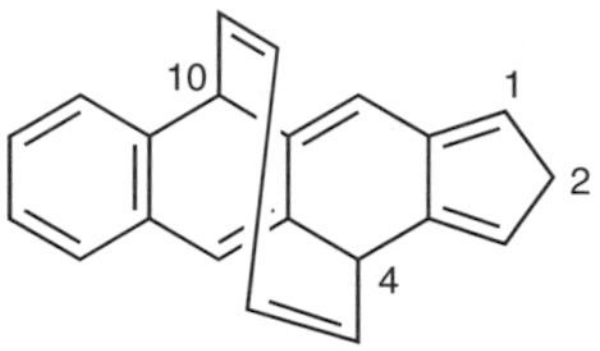

4,10-Dihydro-2H-4,10-
buta[1,3]dienocyclo-
penta[b]anthracene

Other multivalent bridges are treated in the same way as the corresponding substituent groups.

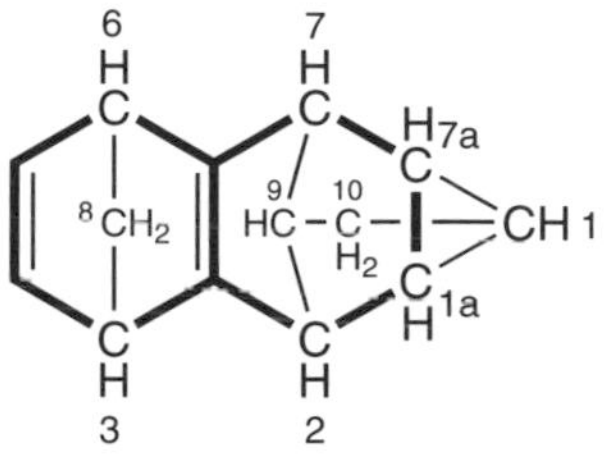

1a,2,3,6,7,7a-Hexahydro-1,2,7-[1]
ethanyl[2]ylidene-3,6-methano-1H-
cyclopropa[b]naphthalene
better: ... -2,7,1-(ethane[1,1,2]triyl)-3,6- ...

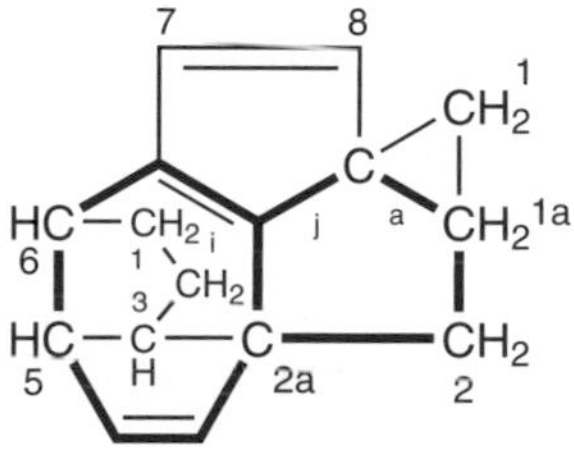

1a,2,5,6-Tetrahydro-
6,2a,5-[1]propanyl
[3]ylidene-1H-cyclo-
penta[ij]cyclo-
propa[a]azulene
better: ... -5,2a,6-
(propane[1,1,3]triyl)-
1H- ...

To name ring systems of Table 1 as bridges their name-ending ...ene is
converted into ...eno.

9,10-Dihydro-9,10-[1,2]benzenoanthracene

trivial name = Triptycene

6b,12b-[1,8]Naphthalenoacenaphtho[1,2-*a*]acenaphthylene

Simple monocyclic hydrocarbon bridges are named by attaching the morpheme **...epi...** in front of the corresponding fusion term.

16*H*-4a,12a-[1,2]Epicyclopentaazuleno[5,6-*b*]anthracene
Chem. Abstr. uses here: ... [1′,2′]-*endo*-cyclo ...

1.2.1.2.3
Spirocyclic Hydrocarbon Systems

A spiro junction exists when two rings are connected through a single (carbon) center. Monospirocycles containing only two aliphatic rings are specified by the term spiro in front of the name of the corresponding acyclic system with the same number of carbon atoms. To discern possible isomers, a numerical descriptor enclosing in square brackets the sums of C atoms present in each branch is placed between the term spiro and the rest of the name. This could be construed as inconsistent because, contrary to the rules for bridged systems, the smaller branch is cited first. If double bonds are present they are given lowest possible locants as usual.

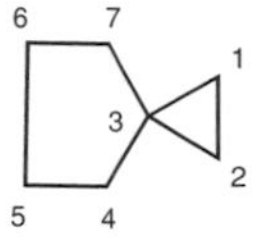

Spiro[2.4]heptane 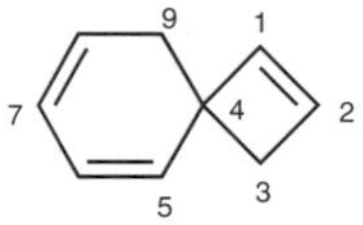   Spiro[3.5]nona-1,5,7-triene

Linear polyspiro compounds of this kind are identified as **dispiro-, trispiro-** etc. -alkanes; here too the numerical descriptors are obtained by enclosing the sums of atoms between the spiro centers in square brackets and separating them by full stops. The smaller peripheral ring is numbered first, and smaller numbers are preferred for the internal path lengths. This leads automatically to smallest possible locants for spiro centers when overall numbering starts on the smaller outer ring at a C atom next to the first spiro junction.

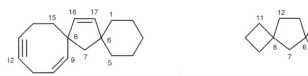

Dispiro[5.1.7.2]heptadeca-9,16-   Trispiro[2.2.1.3.2.2]
dien-12-yne   pentadecane

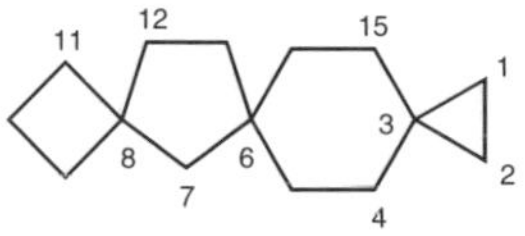

To permit their distinction from related linear isomers the descriptors of **branched spiro compounds** must be supplemented at least by the locants of the **branching spiro centers** which are added as superscripts to the corresponding branching denominators.

Trispiro[2.2.2^6.2.2^{11}.2^3]pentadecane

The former name trispiro[2.2.2.2.2.2]pentadecane was identical with that of the linear isomer shown to the right, whose numerical descriptor can also be made more accurate by using the format ... [2.2.2.2^9.2^6.2^3] ... .

This procedure can easily be extended to all conceivable combinations of branched and linear polyspiro systems.

Hexaspiro[2.0.2.0.2^8.2.2^{13}.0^7.2^4.0.2^{18}.2^3]icosane

In spiro compounds containing at least one fused or bridged (von Baeyer) hydrocarbon system the specifier **spiro** is followed, enclosed in square brackets, by the names of the components including their individual numberings. Spiro positions are again given the lowest possible locants and placed between the component names.

Spiro[cyclopropane-1,9'-fluorene]   Spiro[fluorene-9,1'-indene]

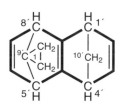

Spiro[cyclopropane-1,9'-[1,4,5,8]tetrahydro [1,4:5,8]dimethanonaphthalene]

For more complex systems, particularly those involving manifest or latent **indicated hydrogen(s)**, some very specific naming procedures are in use which are only vaguely hinted at in the codified IUPAC rules. Since they are in fact applied more or less consistently in the **Ring Index** and by **Chemical Abstracts** a brief illustration is given here.

If a partial structure appearing in a spiro compound already posseses an indicated hydrogen, this hydrogen is placed directly in front of the component name, preceded by its locant and enclosed in square breckets, if necessary. If, however, spiro junction is only possible after a formal dihydrogenation, the additional hydrogen (now called **added hydrogen**), preceded by its locant, is placed in parentheses immediately behind the related spiro locant. Both rules can be valid simultaneously. According to a new proposal only **saturation designators** that are not self-evident should be retained **in front** of the full name.

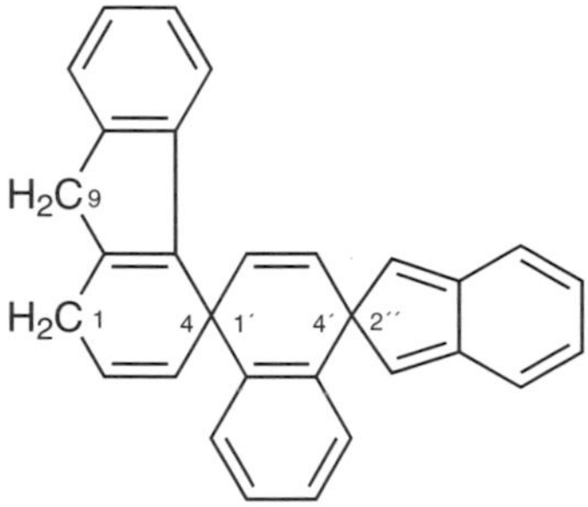

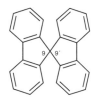

Dispiro[9*H*-fluorene-4(1*II*),1'(4'*H*)-naphthalen-4',2''-(2*H*)indene]
new: 1,9-Dihydro[fluorene-4,1'-naphthalen-4',2''-indene]

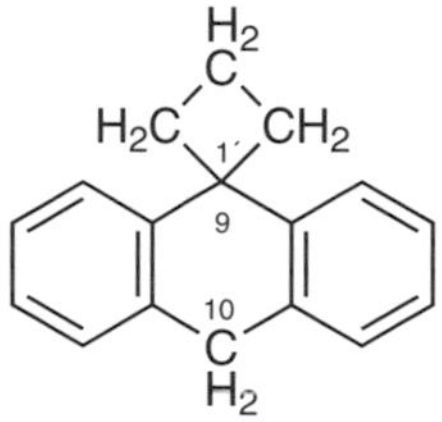

Spiro[anthracene-9(10*H*),1'-cyclobutane]
new: 10*H*-Spiro[anthracene-9,1'-cyclobutane]

If a spiro system is composed of two identical subunits the connectivity locants followed by the term **spirobi..** are placed in front of the name of the partial system.

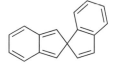

9,9'-Spirobifluorene

1,2'-Spirobiindene

7,7'-Spirobi[bicyclo[4.1.0]hept-3-ene] or 7,7'-Spirobi-[bicyclo[4.1.0]heptane]-3,3'-diene

Di- and polyspiro systems analogously composed of identical subunits can be treated in the same manner.

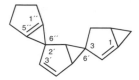

3,6′:2′,6″-Dispiroter[bicyclo[3.1.0]hexa]-1,3′,1″(5″)-triene

1*H*,1′*H*,1″*H*,3′*H*-2,2′:7′,2″-Dispiroter[naphthalene]

Note:
Spiro systems in which the spiro atom is at the same time incorporated into a third ring – i.e., where the spiro junction is **not free** – are not named as spiro compounds but treated as fused polycycles.

9,16b-Dihydrofluoreno[8a,9-*l*]phenanthrene

4a*H*-Benzo[*d*]naphthalene

1.2.1.2.4
Hydrocarbon Ring Systems Linked Through Single or Double Bonds; Ring Assemblies

If two **identical** carbocycles are connected by single or double bonds they are designated as **bi ...yl** or **...ylidene derivatives**. Alternatively, such

assemblies can be named by placing the prefix **bi...** in front of the
unchanged hydrocarbon name.

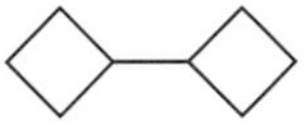

Bicyclobutyl   Biphenyl 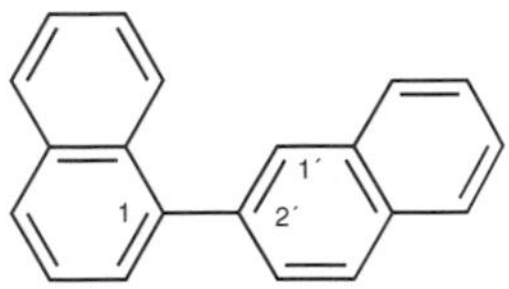   1,2'-Binaphthyl
Bi(cyclobutane)   (1,2'-Binaphthalene)

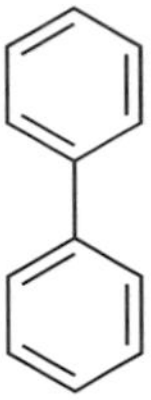

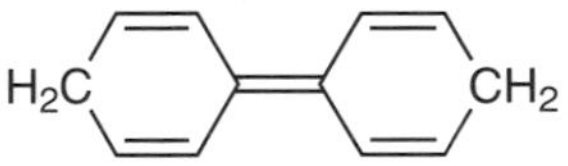

Bi(cyclohexa-2,5-dien-1-ylidene)

If three or more identical cyclic systems are joined together the multiplying prefixes **ter, quater, quinque, sexi, septi, octi, novi,** etc. are attached to
the hydrocarbon (but not the group) name. Members of the polyphenyl
series are treated as exceptions.

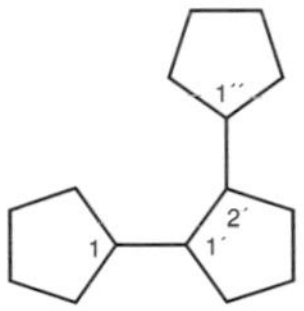

1,1':2',1''-Tercyclopentane 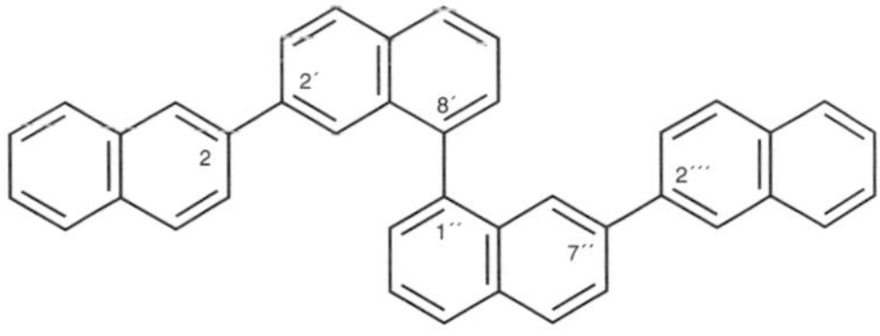   2,2':8',1'':7'',2'''-Quaternaphthalene

p-Terphenyl   o-Terphenyl
(1,1':4',1''-Terphenyl)   (1,1':2',1''-Terphenyl)

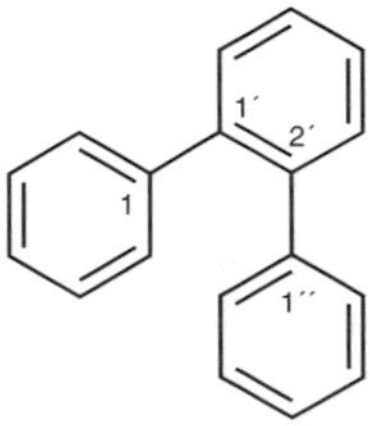

If **non-identical** ring systems are joined together by single or double bonds, one is considered as parent structure and the other(s) as substituent group(s). To select the most senior parent system the following descending order of priorities is obeyed:

a) the highest number of constituting rings,
b) the largest individual ring present,
c) the highest degree of unsaturation,
d) the list of retained trivial names (see Table 1, p. 16).

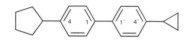

4-Cyclopentyl-4′-cyclopropylbiphenyl

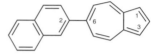

6-(2-Naphthyl)azulene

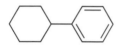

Cyclohexylbenzene

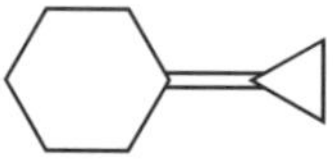

Cyclopropylidenenecyclohexane

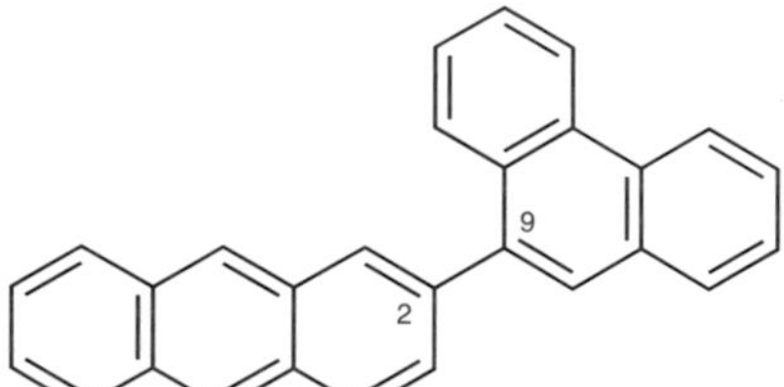

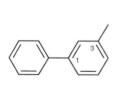

2-(9-Phenanthryl)anthracene

Substituent groups derived from the ring systems treated in this chapter are named as usual:

Biphenyl-3-yl 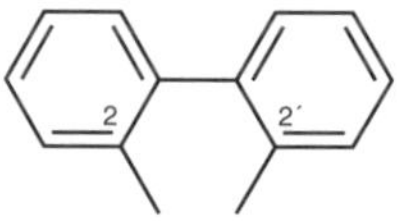   2,2′-Biphenylylene   1,1′:2′,1″-Terphenyl-
                       Biphenyl-2,2′-diyl   2,6′,2″-triyl

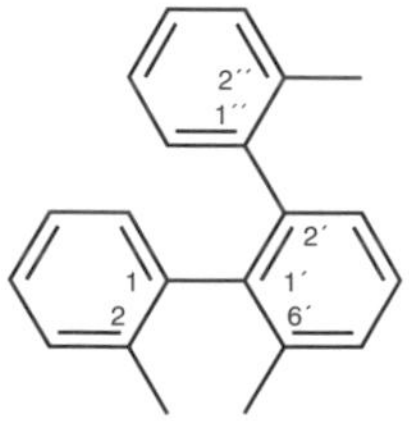

1.2.1.2.5
Cyclic Hydrocarbons with Side Chains

In principle, hydrocarbon systems comprising at the same time rings **and** chains can be treated as ring systems substituted by chains or as chains substituted by rings. Chemical intuition would certainly prefer a central structure with as many substituents as possible and view a smaller unit as substituent of a larger one; in 1972, however, Chem. Abstr. decreed categorically: **rings always have priority over chains!**

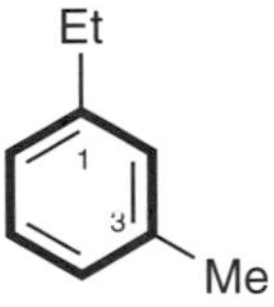

1-Ethyl-3-methylbenzene

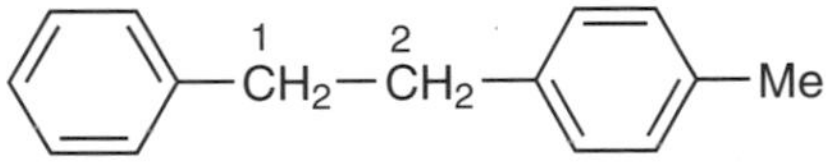

1-Phenyl-2-(4-tolyl)ethane
Chem. Abstr.: 1-Methyl-4-
(2-phenylethyl)benzene

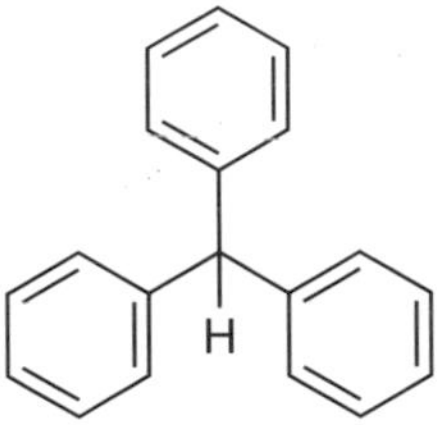

Triphenylmethane
Chem. Abstr.:
Methylidynetrisbenzene

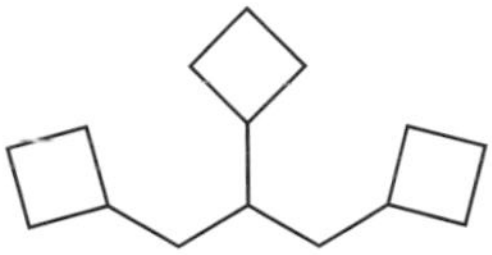

1,2,3-Tri(cyclobutyl)propane
Chem. Abstr.:
Propane-1,2,3-triyltriscyclobutane

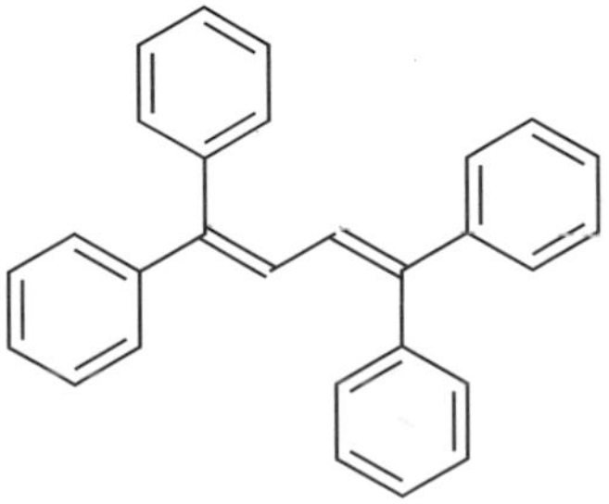

1,1,4,4-Tetraphenylbuta-1,3-diene
Chem. Abstr. since 1972:
Buta-1,3-diene-1,4-bisylidenetetrakis-
benzene
A name conforming better with the
rules would be:
Buta-1,3-diene-1,1,4,4-tetrayltetrakis-
benzene

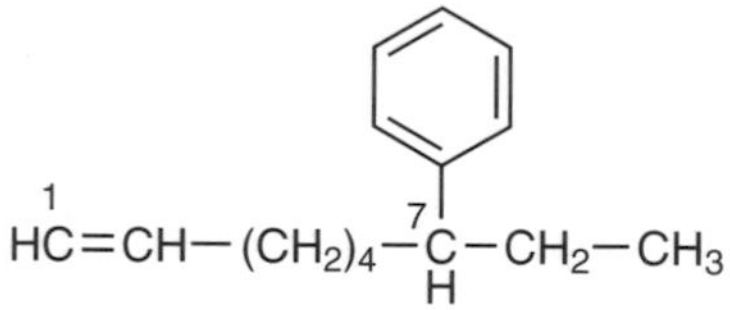

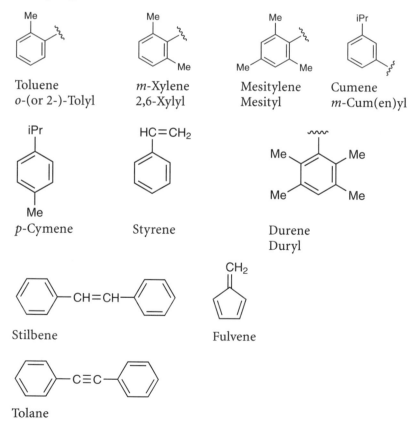

7-Phenylnonene
Chem: Abstr.:
(1-Ethylhept-6-enyl)benzene

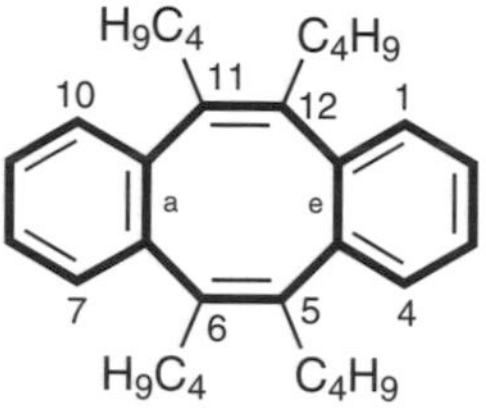

5,6,11,12-Tetrabutyldibenzo[a,e]cyclooctene
(... dibenzo[a,e][8]annulene)

For many aromatic representatives of this series, trivial names are retained which are listed below together with the names of the respective substituent groups.

Toluene
o-(or 2-)-Tolyl

m-Xylene
2,6-Xylyl

Mesitylene
Mesityl

Cumene
m-Cum(en)yl

p-Cymene

Styrene

Durene
Duryl

Stilbene

Fulvene

Tolane

If these trivial systems are further substituted, however, they are treated as substituted benzenes.

1,4-Divinylbenzene
or *p*-Divinylbenzene
not *p*-Vinyl**styrene**

1,2,3-Trimethylbenzene
not Methyl**xylene**
not Dimethyl**toluene**

1,2-Dimethyl-
3-propylbenzene
(not 3-Propyl-
o-**xylene**)

In addition to the trivial names already mentioned, a number of more specific group names for arenes with side chains are also retained.

Benzyl $\quad$ $C_6H_5\overset{\alpha}{-CH_2-}$ $\qquad$ Benzhydryl $\quad$ $(C_6H_5)_2\overset{\alpha}{CH-}$

Phenethyl $\quad$ $C_6H_5\overset{\beta}{-CH_2}\overset{\alpha}{-CH_2-}$ $\qquad$ Styryl $\quad$ $C_6H_5\overset{\beta}{-CH}=\overset{\alpha}{CH-}$

Trityl $\quad$ $(C_6H_5)_3C-$ $\qquad$ Cinnamyl $\quad$ $C_6H_5\overset{\gamma}{-CH}=\overset{\beta}{CH}\overset{\alpha}{-CH_2-}$

Names of polyvalent groups derived therefrom are formed as usual:

Benzylidene $C_6H_5-CH=$ $\qquad$ Cinnamylidyne $\quad$ $C_6H_5-CH=CH-C\equiv$
(traditionally also: Benzal)

1.2.2
Heterocyclic Systems

1.2.2.1
Trivial Names

There are two reasons why the nomenclature of heterocycles is particularly complex. First of all, the **IUPAC rules** are based on two alternative types of nomenclature which are not very clearly delimited, either in the Ring Index or in Chemical Abstracts. Secondly, many more trivial names are retained for heterocyclic compounds than elsewhere. A representative selection of trivial names, particularly of those widely used as parent

Table 2. The most important heterocycles for which trivial names are retained[a]

Thiophene (2-Thienyl) Analogously: Selenophene Tellurophene	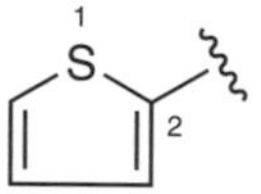
Furan (3-Furyl)	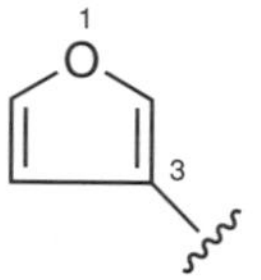
2*H*-Pyran 4*H*-Pyran Analogously: Thiopyran etc.	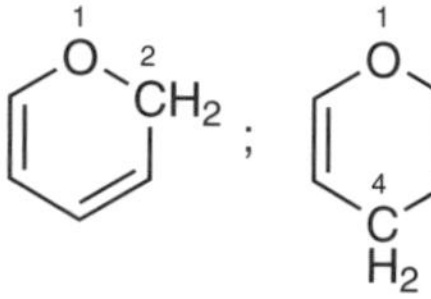
Benzofuran Now: 1-Benzofuran	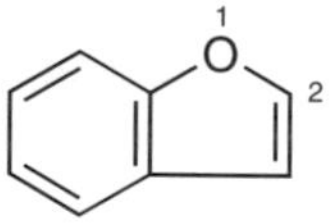
Isobenzofuran Now: 2-Benzofuran	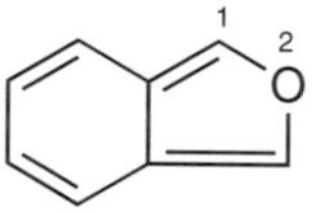
2*H*-Chromene (2*H*-Chromen-3-yl) Analogously: Thiochromene	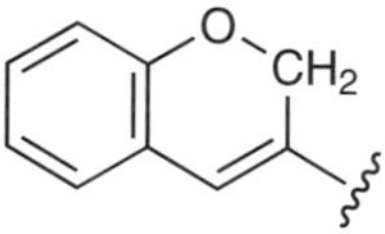
4a*H*-Isochromene (4a*H*-Isochromen-3-yl) Analogously: Isothiochromene etc.	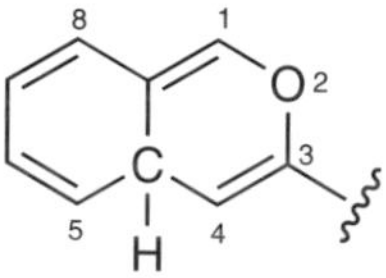
Thianthrene Analogously: Oxanthrene etc. Phosphanthrene etc. Boranthrene etc. Silanthrene etc.	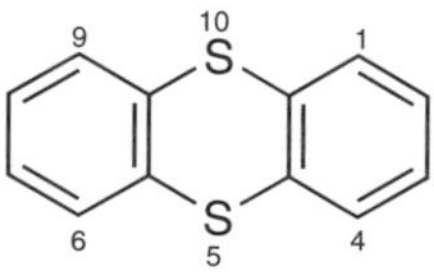

Table 2 (continued)

Xanthene[b] Analogously: Thioxanthene etc.	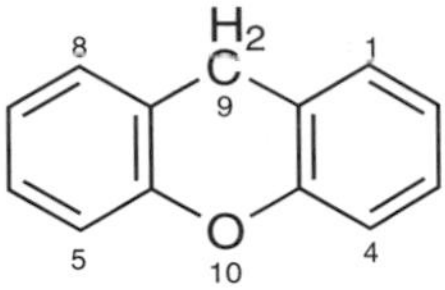
Phenoxathiine Analogously: Phenoxaselenine etc. Phenoxaphosphine etc.	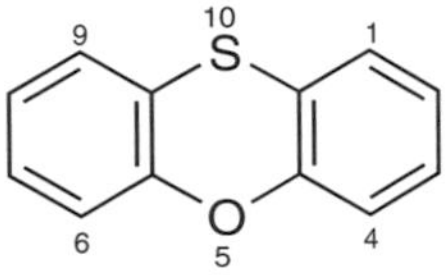
Pyrrole 2H-Pyrrole	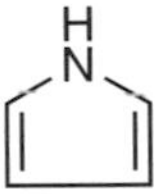 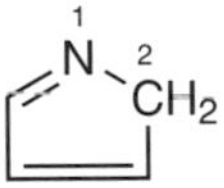
Imidazole 4H-Imidazole	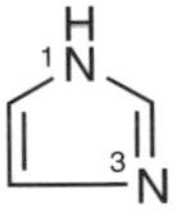 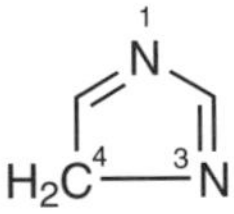
Pyrazole 1H-Pyrazole	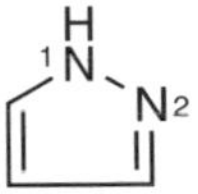 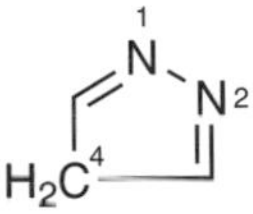
5H-Indazole Indazole	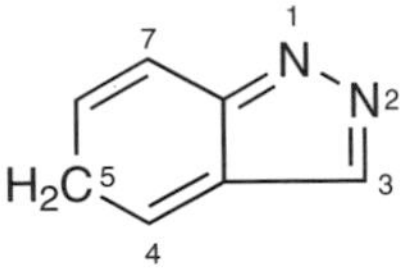 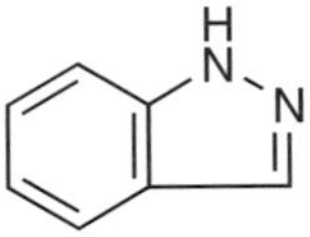
Purine[b]	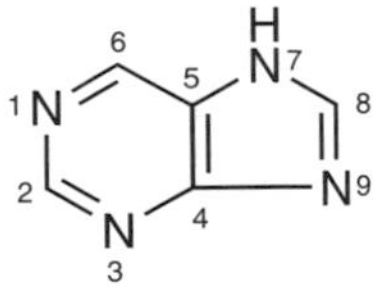
Pyridine (4-Pyridyl)	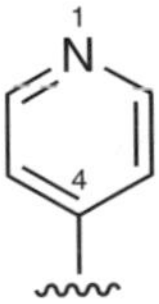

Table 2 (continued)

Pyrazine	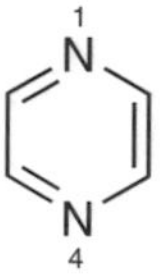
Pyrimidine	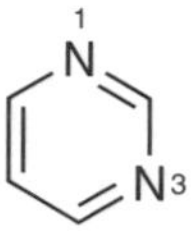
Pyridazine	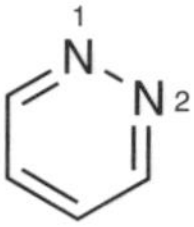
4*H*-Quinolizine[c] Analogously: Phosphinolizine etc.	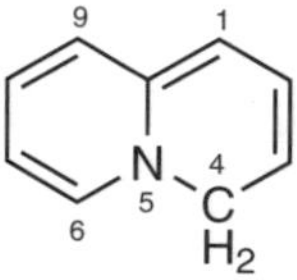
Isoquinoline[c] (3-Isoquinolyl) Quinoline[c]	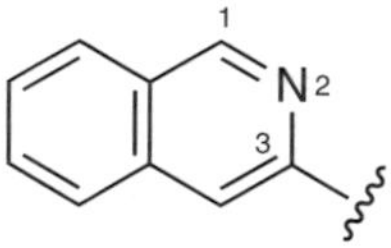 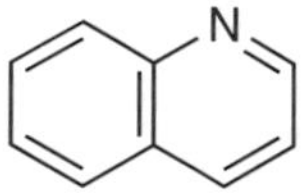
Phthalazine	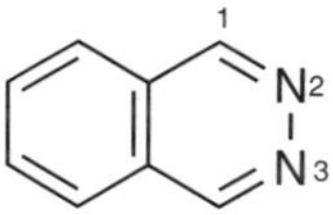
1,8-Naphthyridine Generally: m,n-Naphthyridine	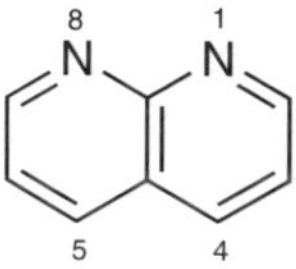
Quinoxaline[c]	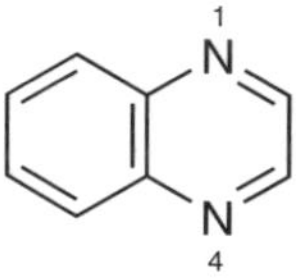

Table 2 (continued)

Quinazoline[c]	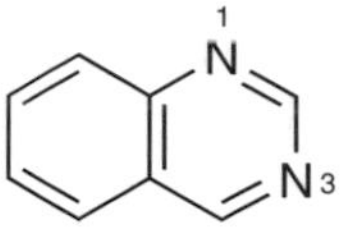
Cinnoline	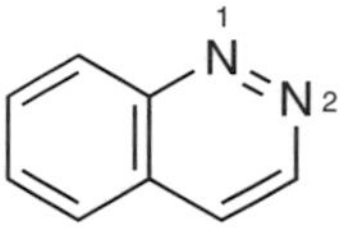
Pteridine	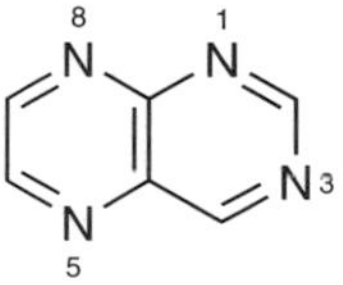
Indolizine Analogously: Phosphindolizine etc.	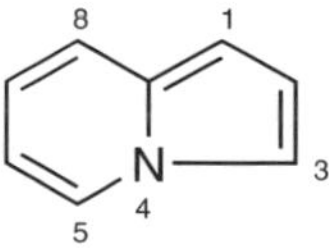
3aH-Indole Indole	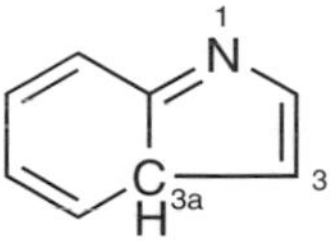 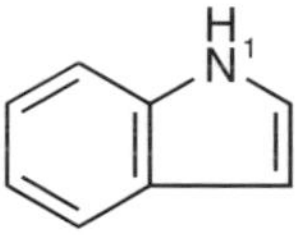
3aH-Isoindole Isoindole	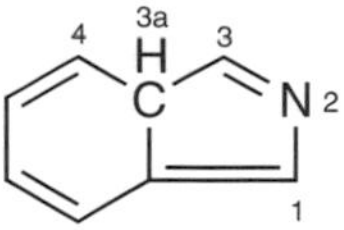 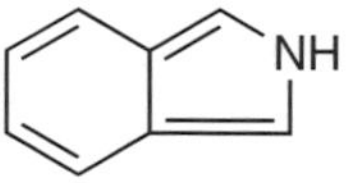
Carbazole[b]	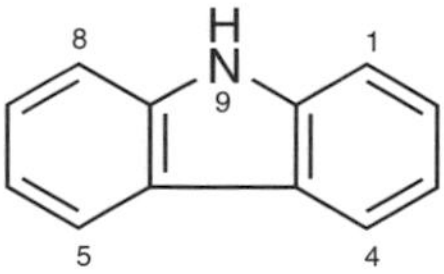
8aH-Carbazole[b]	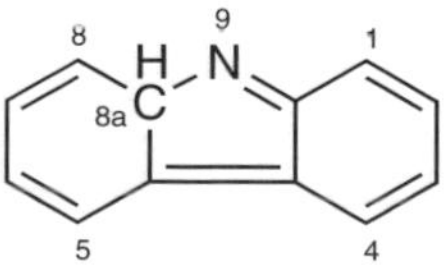

Table 2 (continued)

β-Carboline	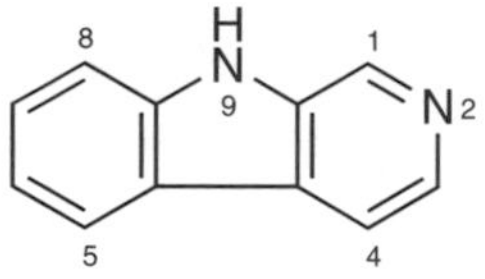
Acridine[b] Likewise: Acridophosphine, Acridarsine (system. numbering)	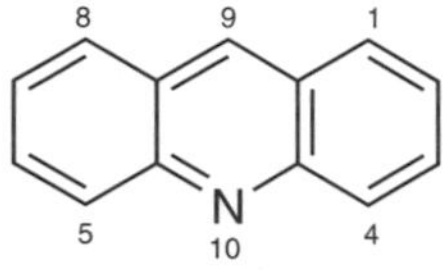
Phenanthridine	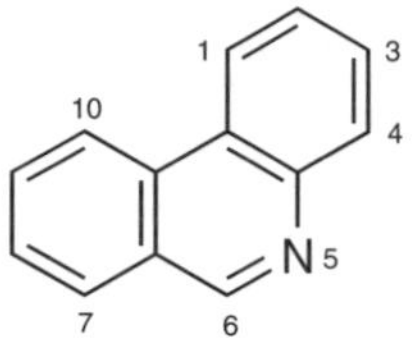
Phenazine	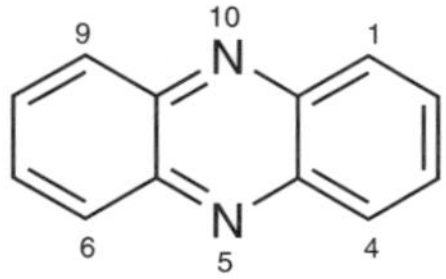
1,7-Phenanthroline Generally: m,n-Phenanthroline	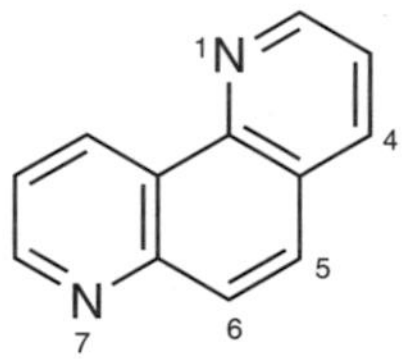
Perimidine	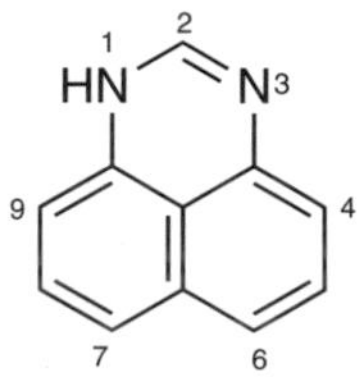
Phenarsazine Analogously: Phenophosphazine etc.	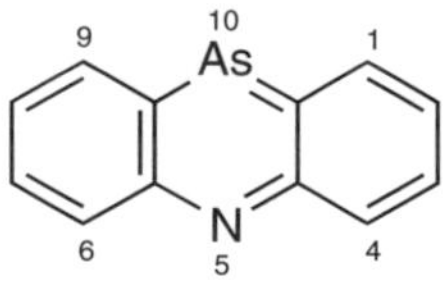

Table 2 (continued)

Phenoxazine	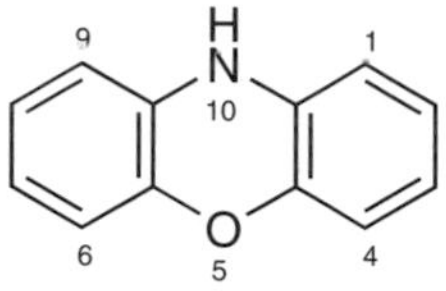
Phenothiazine Analogously: Phenoselenazine etc.	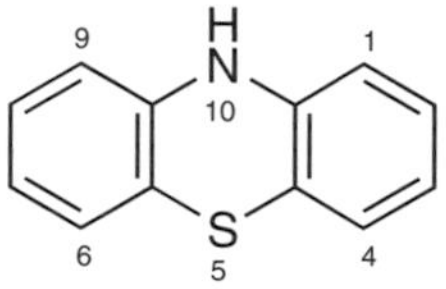
Phosphinoline Likewise: Arsinoline, Isophosphinoline etc.	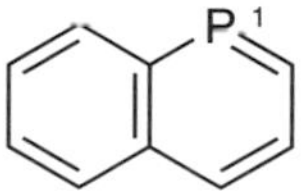
Phosphindole Likewise: Arsindole, Isophosphindole etc.	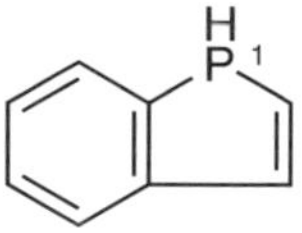
Oxazole Isoxazole New: 1,3- resp. 1,2-Oxazole	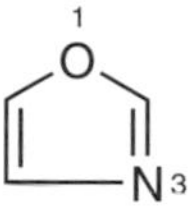 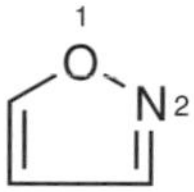
Thiazole Isothiazole New: 1,3- und 1,2-Thiazole	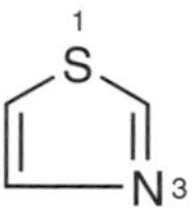 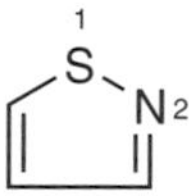
Furazan	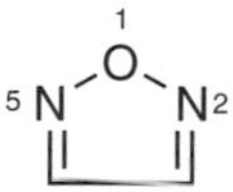

[a] Other trivial names which might be retained can be found in the Ring Index and Chem. Abstr. registers.
[b] Exceptions from systematic numbering.
[c] In German these names still begin with **Ch...**

Table 3. Trivial names of hydrogenated heterocycles

Isochroman Chroman	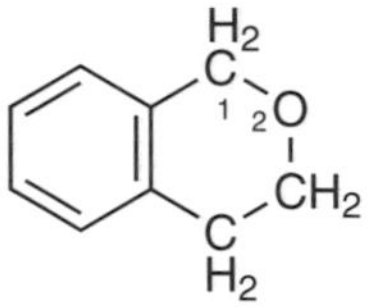 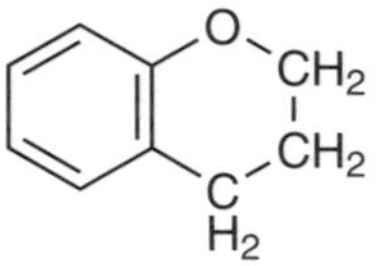
Piperidine (3-Piperidyl)[a]	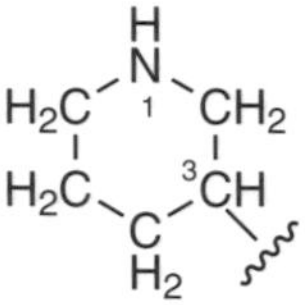
Pyrrolidine	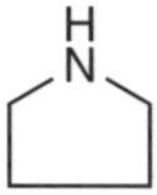
Piperazine (Piperazin-1-yl)	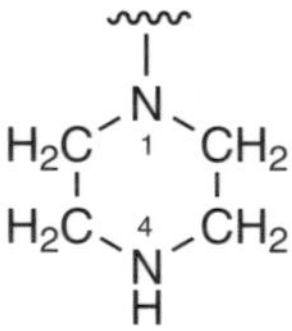
Indoline Isoindoline	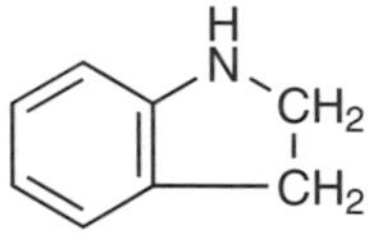 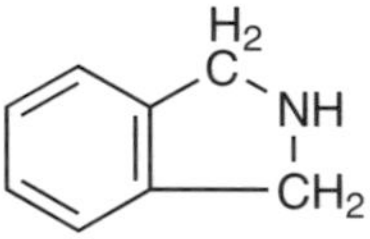
Imidazolidine	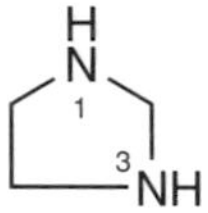
Quinuclidine[b]	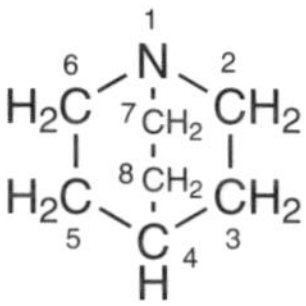
Pyrazolidine	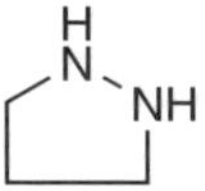

Table 3 (continued)

Morpholine (Morpholin-2-yl)[c]	
Phosphindoline	

[a] 1-Piperidyl = Piperidino
[b] German: Chinuclidin
[c] 4-Morpholinyl = Morpholino

component names for more complex fused heterocycles is provided by Tables 2 and 3. As a rule, the names of the pertinent substituent groups are formed by suffixing the syllable ...**yl** to the name of the heterocycle. Exceptions are marked separately. Numbering generally follows that of related hydrocarbons; if afterwards there is still a choice, hetero atoms are given lowest locants possible. Contrary to the practice for fused hydrocarbons, hetero atoms at fusion positions (that is, belonging to two ore more rings) are also numbered sequentially (and not with a, b, c, etc.). For further details see p. 61.

Use of the trivial names of Table 3 is recommended only for the componds themselves and for the formation of spiro names but not for the construction of fusion names.

Monocyclic hetero systems not having a retained trivial name can basically be named in two different ways. It should be kept in mind, however, that the principles according to which indexing and abstracting organisations opt for one or the other possibility are sometimes anything but obvious.

1.2.2.2
Replacement Nomenclature ("a" Nomenclature)

The simplest method for naming heterocyclic compounds is offered by replacement nomenclature, also known as "a" nomenclature. Accordingly, the name for the underlying hydrocarbon is supplemented by "a"-term prefixes which identify the hetero atoms present. The following Table 4 lists, in descending order of priority, "a" terms for the hetero atoms encountered most frequently in heterocycles. A number of important replacement terms for ions – which can easily be generalized and are subordinate to those of their neutral counterparts – are also shown. If valences other than the standard ones exemplified in Table 4 are to be indicated, this is symbolized by the Greek letter λ with the appropriate bonding number n as superscript, thus leading to the format λ^n.

Table 4. "a" terms for heteroatoms in descending order of priority (incomplete[a])

Element	"a" term	Element	"a" term
$-\overset{..}{\underset{..}{I}}\diagdown$, $-\overset{..}{I}\overset{\oplus}{-}$	λ^3-Ioda, Iodonia	$-\overset{..}{Bi}\diagdown$, $\diagup\overset{\oplus}{Bi}\diagdown$	Bisma, Bismutonia
$-\overset{..}{\underset{..}{O}}-$, $-\overset{\oplus}{\underset{..}{O}}\diagdown$	Oxa, Oxonia	$-Si-$	Sila
$-\overset{..}{\underset{..}{S}}-$, $-\overset{\oplus}{\underset{..}{S}}\diagdown$	Thia, Thionia	$-Ge-$	Germa
$-\overset{..}{\underset{..}{Se}}-$, $-\overset{\oplus}{\underset{..}{Se}}\diagup$	Selena, Selenonia	$-Sn-$	Stanna
$-\overset{..}{\underset{..}{Te}}-$, $-\overset{\oplus}{\underset{..}{Te}}\diagdown$	Tellura, Telluronia	$-Pb-$	Plumba
$-\overset{..}{N}\diagdown$, $\diagup\overset{\oplus}{N}\diagdown$	Aza, Azonia	$-B\diagdown$, $\diagup\overset{\ominus}{B}\diagdown$	Bora, Borata (system.: Boranuida)
$\diagup\overset{\oplus}{N}:$, $:\overset{\ominus}{N}:$	Azanylia, Azanida	$-\overset{\oplus}{B}-$	Boranylia
$-\overset{..}{P}\diagdown$, $\diagup\overset{\oplus}{P}\diagdown$	Phospha, Phosphonia	$-Al\diagdown$	Alumina
$-\overset{..}{As}\diagdown$, $\diagup\overset{\oplus}{As}\diagdown$	Arsa, Arsonia	$-Zn-$	Zinca
$-\overset{..}{Sb}\diagdown$, $\diagup\overset{\oplus}{Sb}\diagdown$	Stiba, Stibonia	$-Hg-$	Mercura

[a] The complete list of "a" prefixes for replacement names is to be found in the Appendix (Table 20)

This treatment is consistently applied only for heteromonocyclic systems with more than 10 ring members, for simple (not fused) bridged, and for simple (not fused) spiro heterocycles. In the domain of smaller heterocyclic systems generally the **Hantsch-Widman** nomenclature system (see next chapter) is used, even in the case of fully saturated compounds. It is only for silicon containing compounds that Chem. Abstr. inexplicably uses exclusively replacement names! If there are more than one hetero atoms present, priority for numbering is first given to an atom of a higher group, within the group of a lower row of the periodic table; only then are all the other hetero elements assigned lowest locants as a set.

Silacyclopropene

Silacyclopentadiene

H_2Si—
HN NH
—SiH_2

1,4-Diaza-2,5-disilacyclo-hexane

7-Oxa-1,5-diaza-8-mercura-bicyclo[4.2.0]octane

1,8-Dioxa-4,11-diaza-cyclotetradecane

3,6-Dioxa-2-thiatetracyclo-[6.3.0.0^{4,11}.0^{5,9}]un-decane

7,14,15-Trithia-3,11-diaza-dispiro[5.1.5.2]pentadecane

Spiro[3-azabicyclo[3.1.0]hexane-6,1'-cyclobutane]

Spiro[cyclohexane-1,4'-[3,7]-dioxabicyclo[4.1.0]heptane]

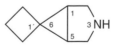

1,6-Dioxa-4-aza-9-azonia-5-borata-spiro[4.4]nonane

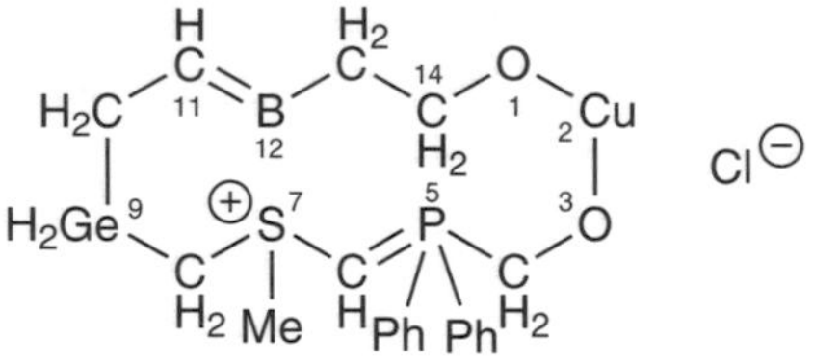

7-Methyl-5,5-diphenyl-1,3-dioxa-7-thionia-5λ^5-phospha-9-germa-
12-bora-2-cupra-cyclotetradeca-5,11-diene chloride

In the case of acyclic systems, **replacement nomenclature** is particularly
suitable above all for longer chains containing at least four internal hetero
atoms. These are numbered so as to give lowest possible locants to all hetero atoms as a whole and then according to the priority order cited above.
Only then are multiple bonds considered for lowest possible locants.

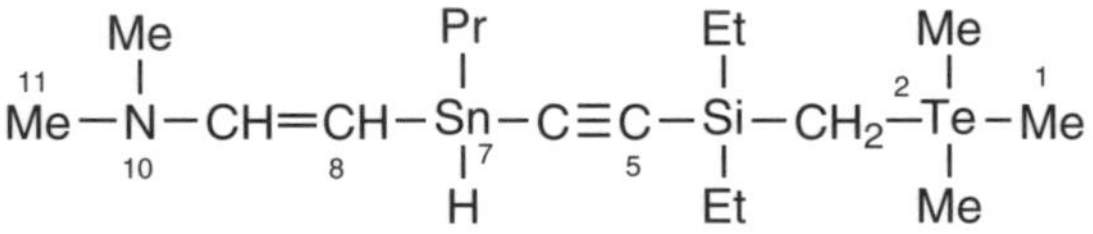

4,4-Diethyl-2,2,10-trimethyl-7-propyl-2λ^4-tellura-10-aza-4-
sila-7-stannaundec-8-en-5-yne

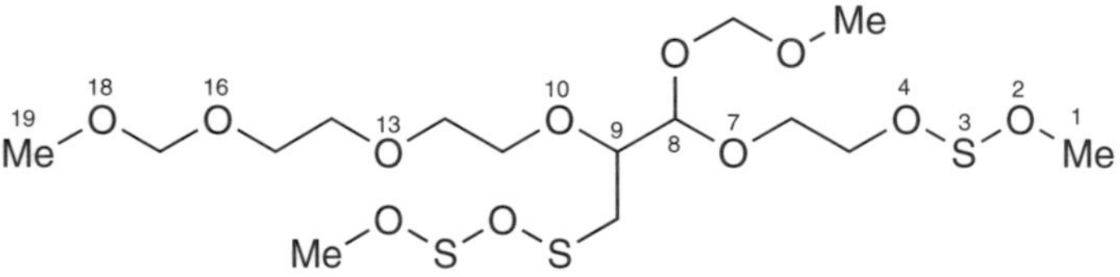

8-(Methoxymethoxy)-9-(3,5-dioxa-2,4-dithiahexyl)-
2,4,7,10,13,16,18-heptaoxa-3-thianonadecane

1.2.2.3
The Hantzsch-Widman System

Heterocyclic compounds with up to ten ring members are generally
named by the extended HW System. In order to do this, a hetero symbol
(or several of these symbols) from the table of "a" terms (Table 4) is combined (eliding the final "a") with a **stem term** of the following Table 5 that
indicates both ring size and degree of saturation. (Unsaturation here
means: maximum number of non-cumulative double bonds – mancude.)

Table 5. Symbols for indicating ring size and saturation degree of heterocycles within the Hantzsch-Widman system. (**A**: O, S, Se, Te, Bi, Hg; **B**: N, Si, Ge, Sn, Pb; **C**: B, F, Cl, Br, I, P, As, Sb)

Ring size	Unsaturated[a]	Saturated[b]	Ring size	Unsaturated[a]	Saturated[b]
3	irene[c]	irane[d]	7	epine	epane
4	ete	etane[d]	8	ocine	ocane
5	ole	olane[d]	9	onine	onane
6A	ine	ane	10	ecine	ecane
6B	ine	inane			
6C	inine	inane			

[a] If at least one double bond and the maximum number of non-cumulated double bonds are present.

[b] When no double bonds are present or possible.

[c] For rings containing (only) nitrogen the traditional ending "irine" can be used.

[d] For nitrogen-containing rings the traditional endings "iridine", "etidine", and "olidine" are still prefered.

The terminal "e" of the above terms is more or less optional (see for example Chem. Abstr.) and could conveniently be dispensed with.

Naming and numbering of hetereo atoms occurs in accordance with the table of "a" terms in such a way that the most senior hetero atom is assigned locant 1; only then are all the other hetero atoms as a set given lowest locants possible (but see p. 61).

2H-Azirine
2H-Azirene

Oxaziridine
Oxazirane

Oxathia-
phosphirane

1H-1,3,2-Di-
azacuprete

1,3,2,4-Diphospha-
diborete

1,2-Dihydro-1,2-
azarsete

1,3,2,4-Diselenadiazetidine
1,3,2,4-Diselenadiazetane

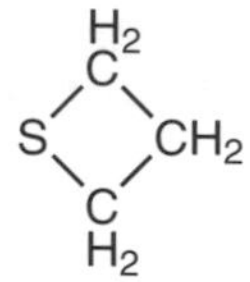

Thietane

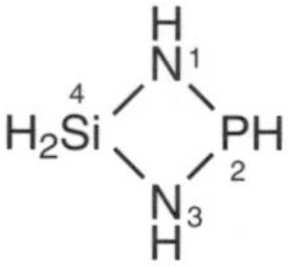

1,3,2,4-Diazaphospha-
siletidine
1,3,2,4-Diazaphospha-
siletane

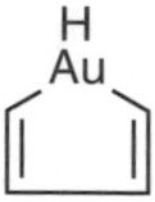

1*H*-Aurole

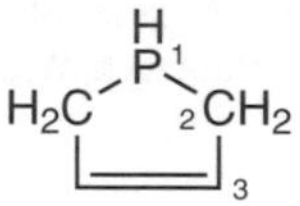

2,5-Dihydro-
1*H*-phosphole

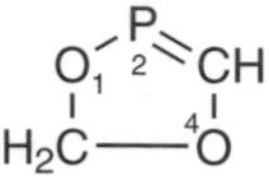

1,4,2-Dioxaphosphole

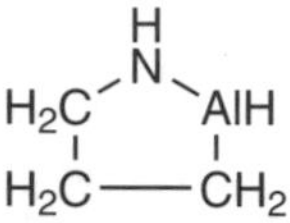

1,2-Azaluminolidine
1,2-Azaluminolane

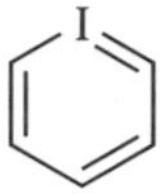

λ^3-Iodinine

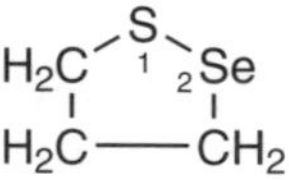

1,2-Thiaselenolane

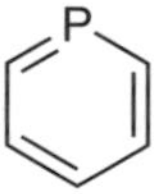

Phosphinine
formerly: Phosphorine

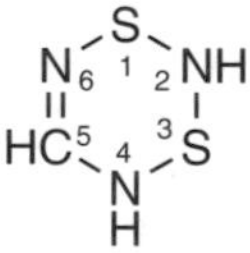

4*H*-1,3,2,4,6-
Dithiatriazine

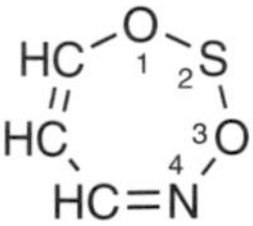

1,3,2,4-Dioxathiazepine

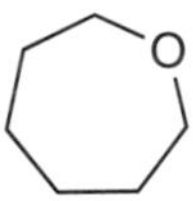

Oxepane

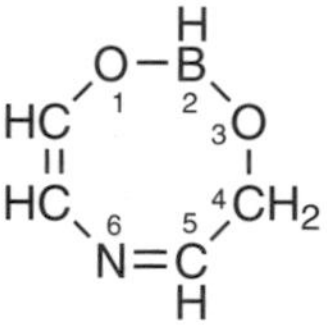

4*H*-1,3,6,2-
Dioxazaborocine

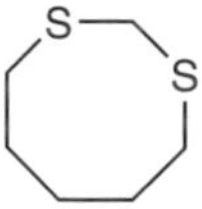

1,3-Dithiocane

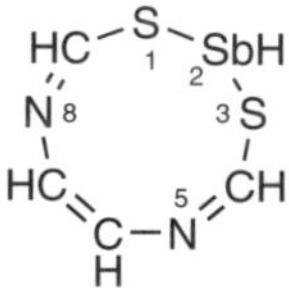

1,3,5,8,2-Dithiadiaza-
stibonine

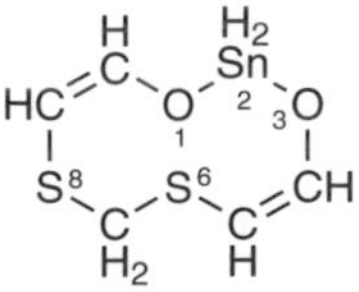

1,3,6,8,2-Dioxa-
dithiastannecine

1.2.2.4
Fused Heterocyclic Systems

Provided that no retained trivial name is applicable, fused heterocyclic systems are given Hantzsch-Widman names. Fusion of components follows the same pattern as established for corresponding hydrocarbons and an intricate series of seniority rules must be run through for selecting the parent component. Traditional abbreviated terms are retained for the following fusion components: **furo, imidazo, (iso)quino, pyrimido, thieno.**

a) In determining the base component, **any heterocyclic** system takes precedence over any carbocyclic entity, however large the latter may be.

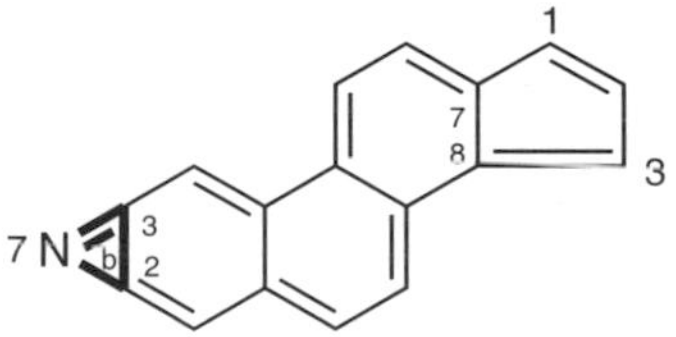

Cyclopenta[7,8]phenanthro[2,3-*b*]azirine

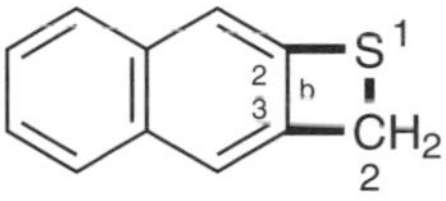

2*H*-Naphtho[2,3-*b*]thiete

This rule is sometimes disregarded by Chem. Abstr. and the Ring Index for oxygen-containing three-membered rings!

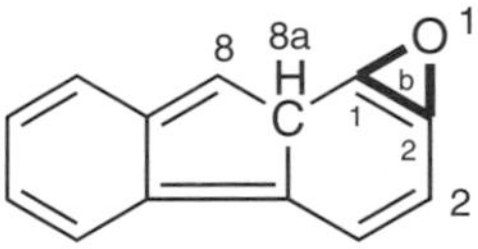

Correct: 8a*H*-Fluoreno[1,2-*b*]oxirene
C. A.: 1,2-Epoxy-8a*H*-fluorene
R. I.: 8a*H*-Oxireno[*a*]fluorene

Related thia derivatives are named as **epithio ...** compounds by both handbooks.

b) The next criterion gives preference to a **component containing nitrogen** (because of the overwhelming significance and sheer abundance of nitrogen cycles).

Pyrano[2,3-c]pyrrole

[1,3]Oxaphosphinino[4,5-*b*]azirine
Formerly: [1,3]Oxaphosphorino[4,5-*b*]azirine

c) If no nitrogen-containing ring is present, priority is given to that component which includes at least one **senior element** according to the table of "a" terms (nothwithstanding ring size and number of other hetero atoms).

[1,2]Phosphaborinino[5,4-*b*]furan

7,8-Dihydro-6*H*-[1,4]dithiino[2,3-*b*]pyran

d) If still no decision is possible, precedence is given to the (trivial) component having the **largest number of rings.**

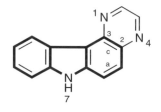

7*H*-Pyrazino[2,3-c]carbazole, not:
7*H*-Indolo[3,2-*f*]quinoxaline

e) The next priority criterion deals with **ring size**. If the most senior element is found in the larger as well as in the smaller ring the larger ring wins as parent component, regardless of the nature and number of other hetero elements present, with the exception of nitrogen, of course (see b).

4H-[1,3,2]Dioxasilolo[4,5-c]pyran

f) If the assignment is still ambiguous the **larger number** of (any) **hetero elements** becomes decisive.

[1,3]Oxaphospholo[4,5-d][1,3,2]oxarsaborole

g) If criterion f) is insufficient, the **greater variety** of hetero atoms becomes valid, if necessary with respect to the "a" term Table 4 (p. 52).

Pyrimido[4,5-b][1,4]azasiline

6H,8H-[1,2,4]Dioxazino[4,3-c][1,3,2]oxazaphosphinine

h) Next, the **greatest number** of senior atoms according to Tables 4, 20 is to be considered.

[1,2,7]Oxadiazepino[4,5-f][1,3,5]dioxazepine

i) If all the preceding criteria fail, that component takes priority whose hetero atoms had **lowest position numbers and/or letters** before fusion.

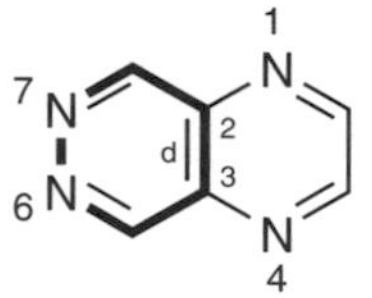

Pyrazino[2,3-*d*]pyridazine

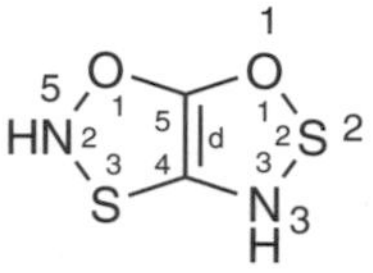

3*H*,5*H*-[1,3,2]Oxathiazolo[4,5-*d*] [1,2,3]oxathiazole

The examples shown above clearly demonstrate that the general fusion principles for heterocyclic systems fully resemble those for analogous hydrocarbon systems. For two-ring systems containing a benzo unit, however, Chemical Abstracts and the Ring Index rigorously apply an exceptional rule which appears only as a subordinate alternative in the IUPAC manuals. Here, only the heteroatoms and not the fusion positions are designated in accordance with the numbering principles outlined below. The names thus formed can be used as (parent) component names for higher fused systems.

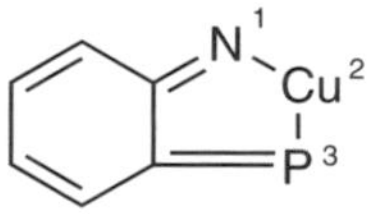

[1,3,2]Benzoazaphosphacuprole
not: Benzo[d][1,3,2]azaphosphacuprole

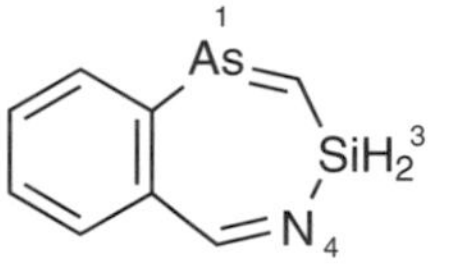

3*H*-[4,1,3]Benzoazarsasilepine
not: 3*H*-Benzo[e][1,4,2]azarsasilepine

If a central system is fused to two (or more) identical peripheral senior components, name construction occurs as shown below:

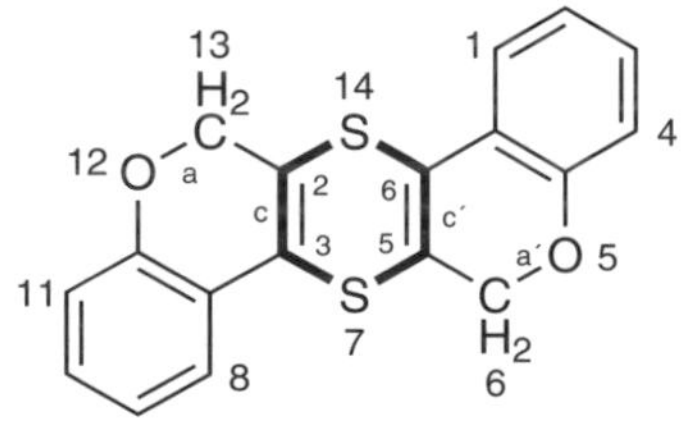

6*H*,13*H*-[1,4]Dithiino[2,3-*c*:5,6*c'*]bis[1]benzopyran

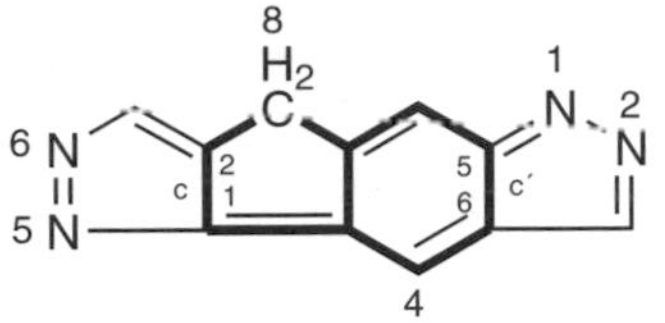

8H-Indeno[1,2-c:5,6-c']dipyrazole

Conclusive **peripheral numbering** of fused heterocyclic systems must first of all pay attention to the generalized orientation rules detailed for corresponding carbocyclics (p. 22). If these are not sufficient, that orientation is chosen which leads to lowest locants by evaluating the following criteria one after another:

a) for hetero atoms as a whole (but see p. 55),
b) for hetero atoms according to the "a" term seniorities (Tables 4, 20),
c) for C atoms common to two or more rings,
d) for hetero atoms of the same element that are not in fusion positions,
e) so that an internal hetero atom lies as close as possible to the lowest numbered fusion position,
f) for indicated hydrogen.

It should be noted once again that hetero atoms in fusion positions are numbered sequentially by plain numbers and taken into account in the names of both components to be fused together.

The following more complex examples may illustrate and clarify these fusion and numbering principles.

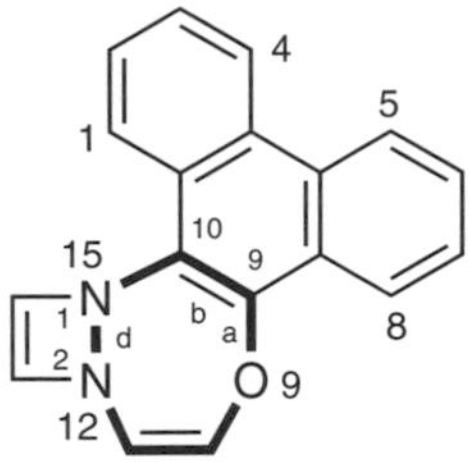

[1,2]Diazeto[1,2-d]phenanthro[9,10-b][1,4,5]oxadiazepine

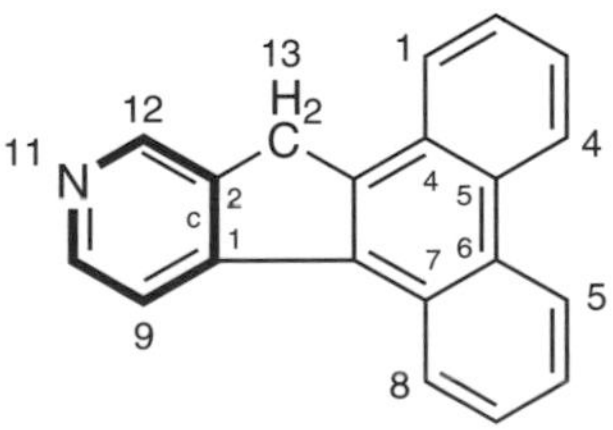

13H-Dibenzo[4,5:6,7]indeno[2,1-c]pyridine

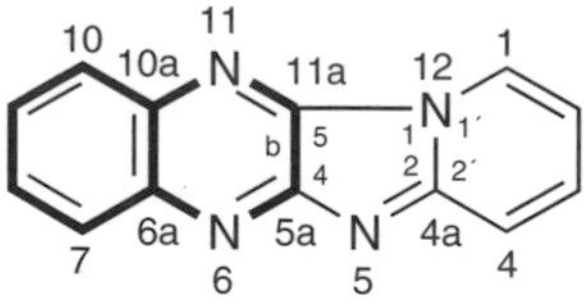

correct 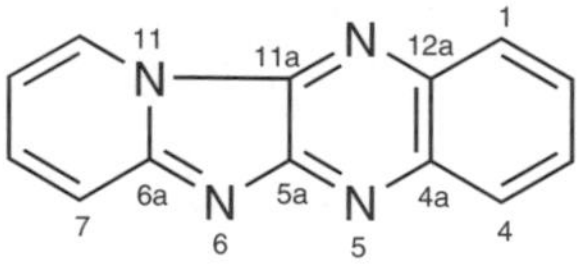   incorrect

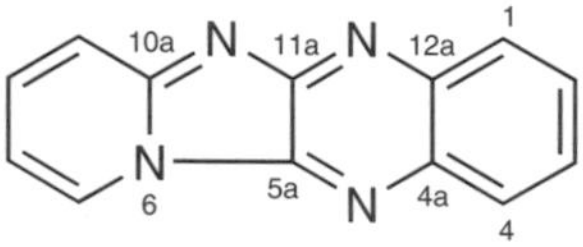

correct 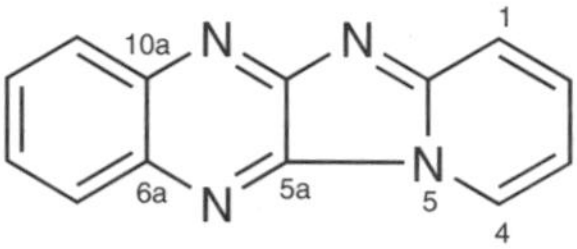   incorrect

Pyrido[1',2':1,2]imidazo[4,5-b]quinoxaline

In the above example the numbering of the nitrogen atoms is always the same; for the fusion locants, however, the following order applies: 4a, 5a, 6a, 10a... < 4a, 5a, 6a, 11a... < 4a, 5a, 10a... < 5a... (criterion c).

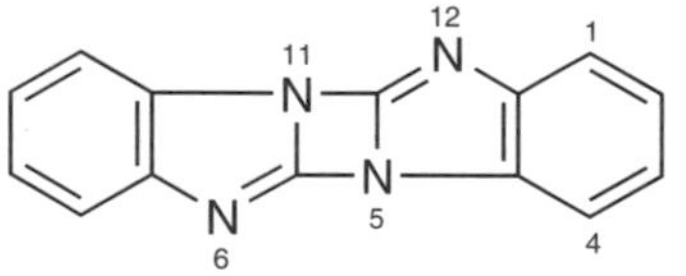

not:

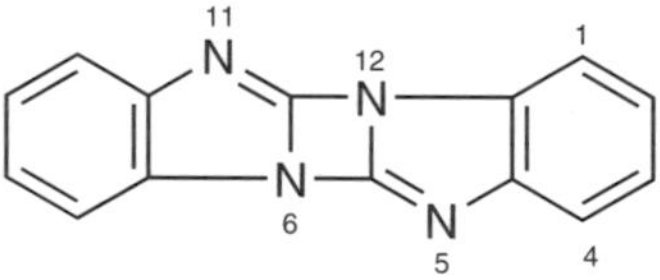

[1,3]Diazeto[1,2-a:3,4-a']di(benzimidazole) (criterion d)

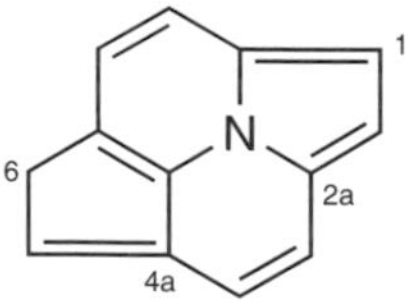

not:

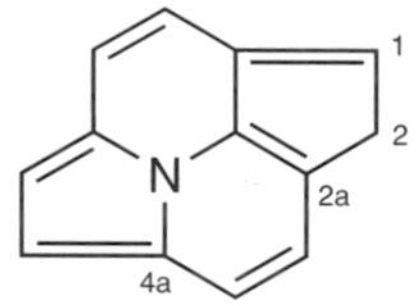

6H-Cyclopenta[de]pyrrolo[2,5-ij]quinolizine (criterion e)

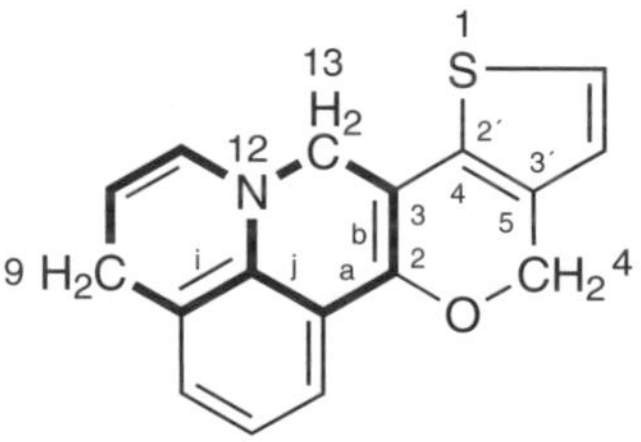

4H,9H,13H-Benzo[ij]thieno[2',3':4,5]pyrano[2,3-b]quinolizine

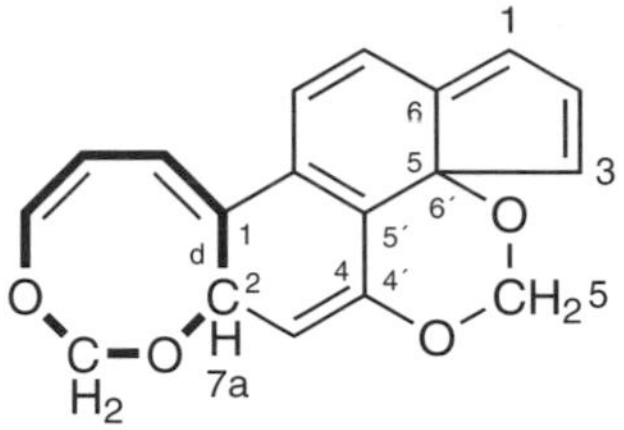

7aH-Cyclopenta[5,6][1,3]dioxino[4',5',
6':4,5]naphtho[2,1-d][1,3]dioxocine

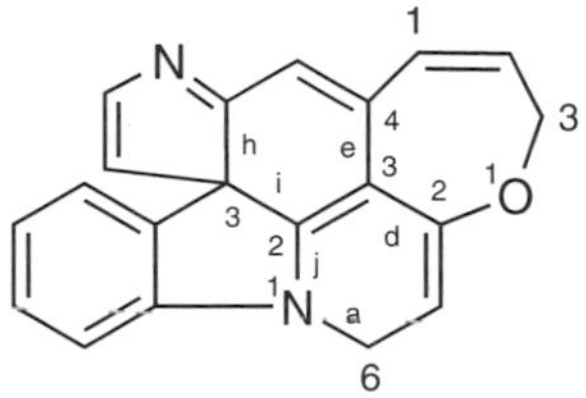

3H,6H-Indolo[3,2,1-ij]oxepino
[2,3,4-de]pyrrolo[2,3-h]quinoline

Actually, in the above example **carbazole** would be the most senior base component; since in that case one of the attached components (**oxepine**) would have to be fused at the same time with the parent component and one of the other attached components (**pyridine**) the second choice (**quinoline**) comes into force.

Cationic systems are provided with the endings -**ium**, -**diium,** etc. (see p. 99).

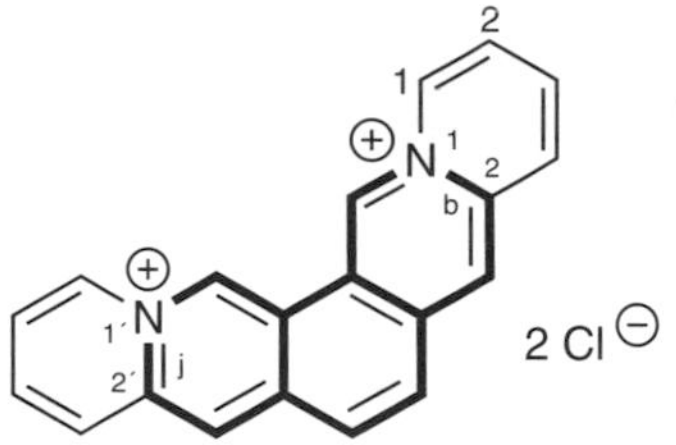

Dipyrido[1,2-b:2',1'-j][2,9]phenanthrolinediium dichloride

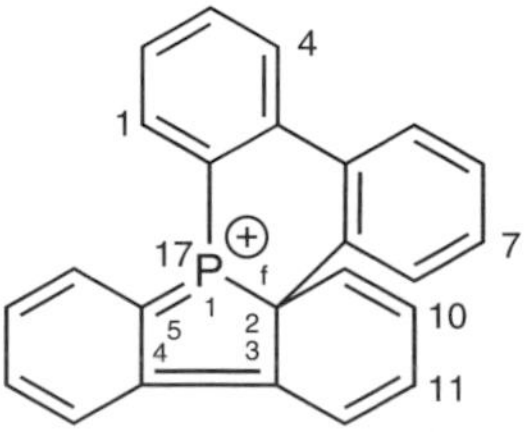

Dibenzo[2,3:4,5]phospholo[1,2-f]phosphanthridinium

When for certain fused heterocyclic systems application of fusion principles is absolutely impossible, **replacement nomenclature** comes to the rescue.

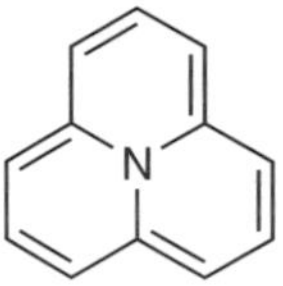

Pyrido[2,1,6-
de]quinolizine

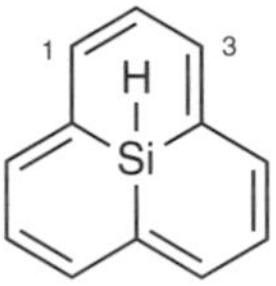

but: 9*bH*-9*b*-Sila-
phenalene

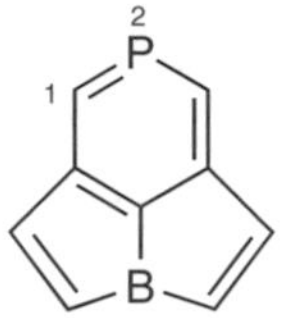

2-Phospha-5*a*-bora-
cyclopenta[*cd*]indene

(Identical) heterocycles linked by single bonds are treated like related carbocycles:

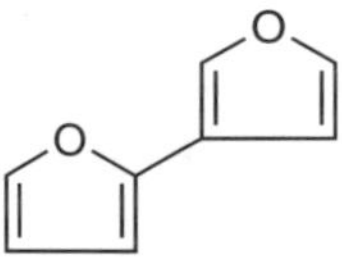

2,3′-Bifuryl
or: 2,3′-Bifuran

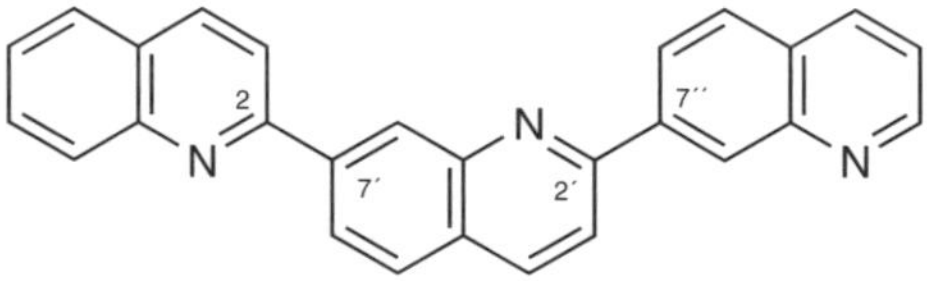

2,7′:2′,7″-Terquinoline

After their components have been arranged according to the preceding rules, the names of **fused heterospirocyclic systems** are assembled in the same way as their carbocyclic counterparts (alphabetical order!).

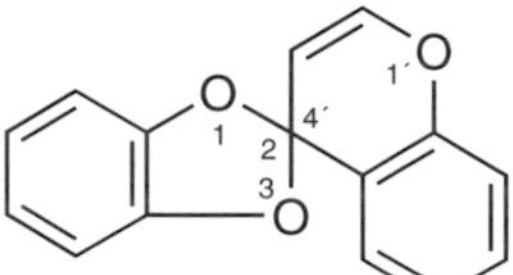

Spiro[1,3-benzodioxole-2,4′-[1]benzopyran]

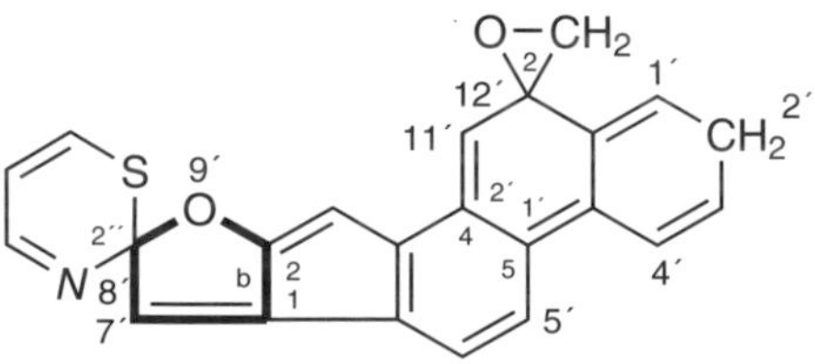

2′*H*-Dispiro[oxirane-2,12′-naphtho[2′,1′:4,5]indeno[2,1-*b*]furan-8′,2″-
[1,3]thiazine]

Dispiro[cyclopenta[1,2-*b*:4,3-*b'*]dipyridine-9,2'-thiirane-3',9''-fluorene]

6,6'-Spirobi[dibenzo[*d,f*][1,3,2]dioxagermepine]

Fused heterocyclic systems containing **additional hydrocarbon bridges** are again treated like their carbocyclic anlogues. If, on the other hand, **hetero bridges** occur in such compounds, they are expressed as composite prefixes to the name in the following way: the front term **ep(i)** is succeeded by the hetero symbols according to the "a"-term table with a terminal "**o**" instead of "a" (which is elided when followed by a vowel). In unambiguous cases ep(i) can be ommitted. (**Beilstein,** however, always uses this front term if it does not switch over to totally systematic formats, as shown in the following examples).

−N=N−NH− Azimino (new: triaz[1]eno), −N=N− Azo (diazeno), −NH−NH− Biimino (diazano)

−NH− Imino (epimino, epiazano), −N= Nitrilo (azeno), −N< Azanetriyl etc., −PH− Phosphano etc., −SiH$_2$− Silano etc.

−O−, −S− Epoxy, epithio (but: −SH$_2$− λ^4-Sulfano etc.); −S−S− Epidithio;

−O−S− Epoxythio; −S−O−NH− Epithioximino;

−Se−O−Te− Episelenoxytelluro;

−O−(CH$_2$)$_n$− Epoxyalkano; −O−(CH$_2$)$_n$−S− Epoxyalkanothio;

−S−(CH$_2$)$_n$−N= Epithioalkanonitrilo;

−(CH$_2$)$_n$−O−(CH$_2$)$_n$− Alkanoxyalkano;

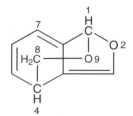

1,4-Dihydro-1,4-(epoxymethano)isobenzofuran
Beilstein: ... oxaethano ...

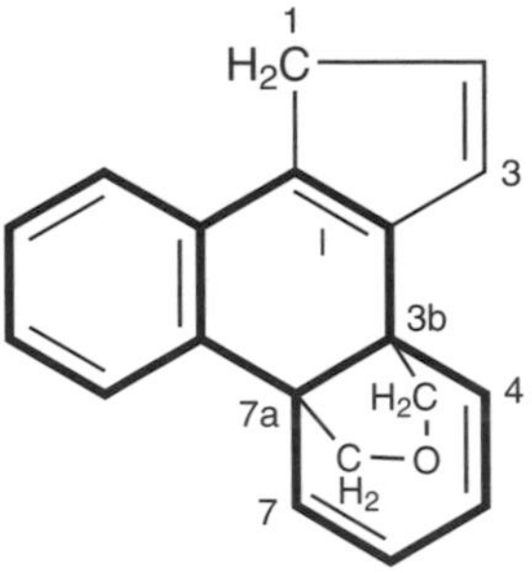

3b,7a-(Methanoxymethano)-1*H*-cyclopenta[*l*]phenanthrene
Beilstein: ... [2]Oxapropano...

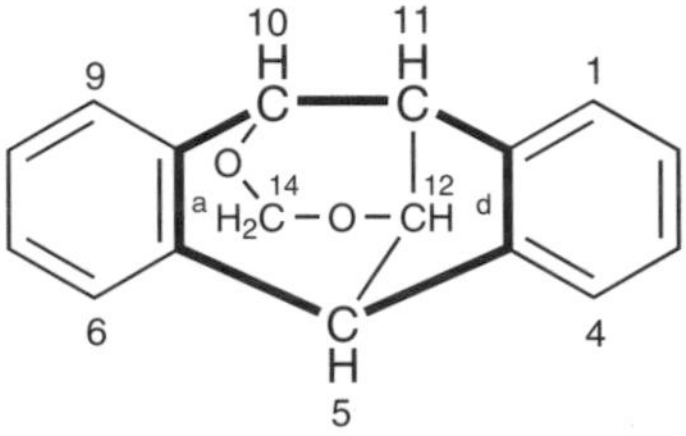

Chem. Abstr.:
10,11-Dihydro-10,5,11-(epoxy-methanoxymetheno)-5*H*-dibenzo[*a,d*]cycloheptene

For the above complex bridge **Beilstein** uses a totally systematic replace-ment name such as [1,3]dioxabutane-1,4,4-triyl; according to a recent **IUPAC** recommendation this would be: epoxymethanoxymethanetriyl.

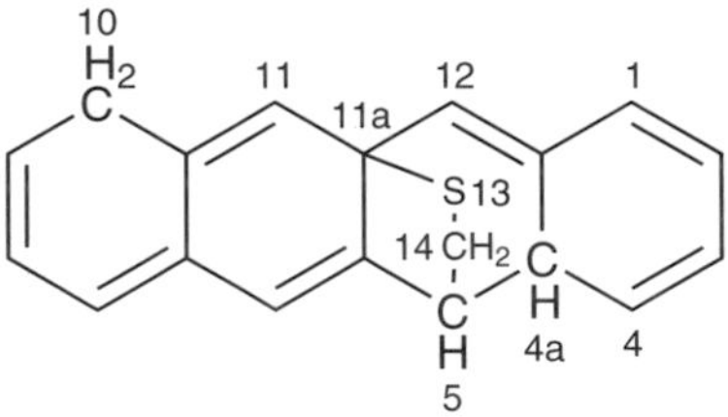

4a,5-Dihydro-11a,5-(epithio-methano)-10*H*-naphthacene
(new: ... tetracene)
Beilstein: ... (thiaethano)...

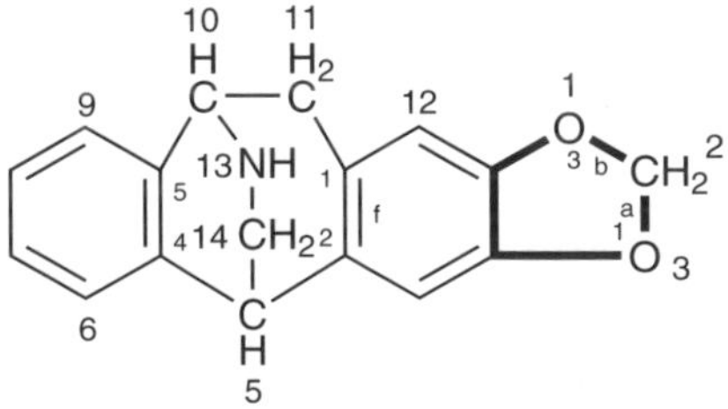

10,11-Dihydro-10,5-(iminome-thano)-5*H*-benzo[4,5]cyclo-hepta[1,2-*f*][1,3]benzodioxole
Beilstein: ... (azaethano)...

If whole heterocyclic systems appear as bridges they are treated like analogous carbocyclic bridges.

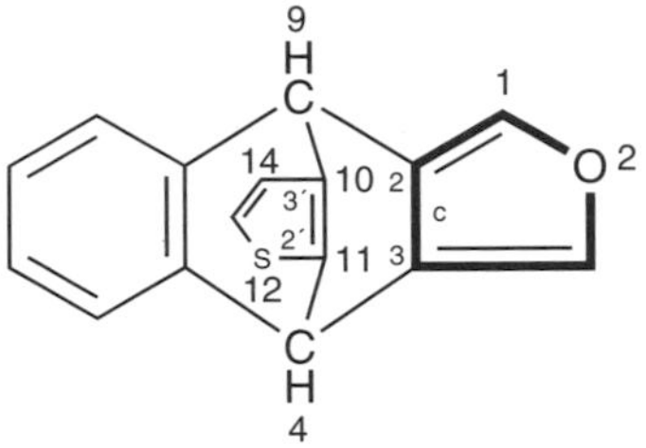

4,9-Dihydro-4,9-[2′,3′]thiopheno-naphtho[2,3-c]furan

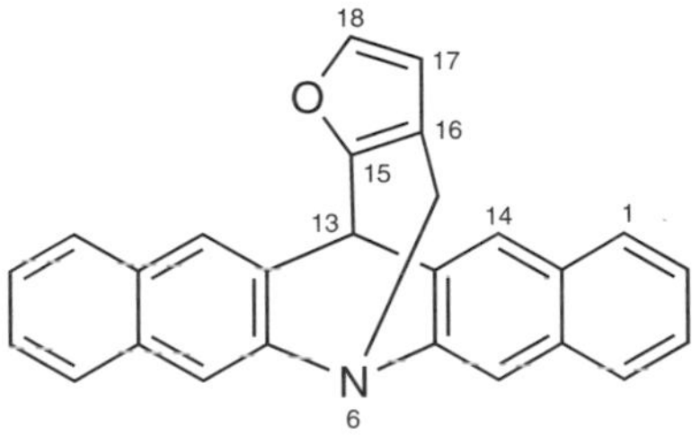

13H-13,6-[2,3]Furanomethano-dibenzo[b,i]acridine

If replacement nomenclature has to be applied for a fused assembly, upon which a bridged system is based, the same holds for the bridges.

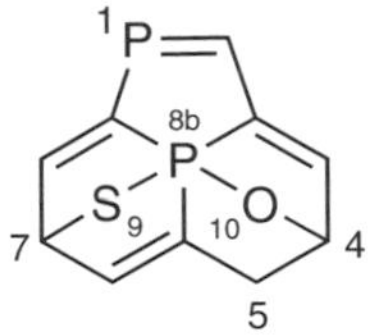

5,7-Dihydro-4H-10-oxa-9-thia-1,8bλ^5-diphospha-7,8b:4,8b-dimethanocyclopenta[de]naphthalene

Concerning the selection of parent structures for bridged systems, special attention must be paid to a really serious discrepency between the current **IUPAC recommendations** and **Chem. Abstr. practice:** according to **IUPAC** the base component should have the **smallest** number of **hetero atoms** while **Chem. Abstr.** gives precedence to the one with the **largest** number.

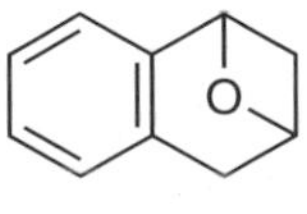

IUPAC: 1,2,3,4-Tetrahydro-1,3-epoxynaphthalene

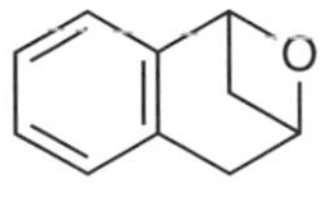

Chem. Abstr.: 3,4-Dihydro-1,3-methano-1H-2-benzopyran

To close this section, attention is drawn to another of the inconsistencies repeatedly found in the indexes of Chem. Abstr. Until 1981 the following frequently cited phosphafluorene system is designated as 5*H*-dibenzophosphole; but starting in 1982 and in compliance with the (semi)trivial name phosphindole, prescribed by Table 2, the same compound is officially designated as 5*H*-benzo[b]phosphindole, although the former name still sporadically reappears.

Until 1981: 5*H*-Dibenzophosphole
(prefered by IUPAC)
Since 1982: 5*H*-Benzo[*b*]phosphindole

1.3
Phane Nomenclature

1.3.1
Cyclophanes

Complex polycyclic compounds where cyclic and catenic subunits are chained together in regular or irregular succession to form a new **super-ring system** are generally called cyclophanes. In principle, all these super-cycles can be named with the existing rules as bridged polycyclic systems as consistently praticed by **Chem. Abstr.** For simple cases the fundamental von Baeyer rules (see p. 29) are sufficient, elsewhere fusion nomenclature (see p. 14) finds application wherever possible; additional connecting units are then treated as additional bridges.

Even for very closely related compounds such procedures frequently lead to utterly different names and – most detrimentally – in many cases to the total disappearance of distinctive cyclic (mostly [het]arenic) subentities. Many efforts have therefore been made to develop a homogeneous nomenclature system for cyclophanes of most divergent appearances. However, none of them, is fully convincing. On the basis of these earlier endeavors the **IUPAC Commission on the Nomenclature of Organic Chemistry (CNOC)** has finally succeeded in presenting a very simple and easily communicable method for naming cyclophanes of all types. Its salient features will be outlined here.

The new system consists essentially in an ingenious adaption of replacement nomenclature, in that the cyclic subunits of a cyclophane are each understood as individual **superatoms** and as such considered equivalent to and sequentially numbered like all other ring atoms. In the final total name, these superatoms are accounted for by **arena, cycloalkana,** etc.

replacement terms which are arranged in compliance with a complexity order similar to that valid for fused systems. The six main criteria of this hitherto non-finalized priority order are to be checked sequentially as follows:

a) heterocycles are always senior to carbocycles,
b) a ring system with the larger number of rings has precedence,
c) hetero atom priority follows the order laid down in the tables of "a" terms,
d) a ring system containing the largest individual ring is preferred,
e) a ring system with the larger number of hetero atoms of any kind ranks first,
f) the least hydrogenated ring system is preferred.

In constructing the cyclophane name, locant numbers and, in parentheses, attachment positions are placed in front of the **cycla** terms. The cyclophane character is expressed by a composite designator at the end of the name that consists of the morphemes **cyclo** and **phane** enclosing a numerator for the members of the superring. Final numbering then assigns lowest locants possible to the most senior superatoms in accordance with the predetermined numbering scheme.

In the following examples the **polycyclic name** reflecting conventional naming procedures is generally given first and then the new IUPAC cyclophane name derived from the associated **supergraph.** For a better understanding, the basic skeletons are drawn in bold print where appropriate. Current trivial or traditional names of some prototypical compounds are also shown.

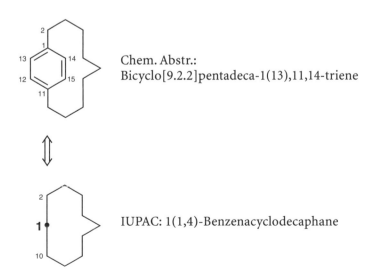

Chem. Abstr.:
Bicyclo[9.2.2]pentadeca-1(13),11,14-triene

IUPAC: 1(1,4)-Benzenacyclodecaphane

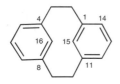

Chem.: Abstr.:
Tricyclo[9.3.1.1⁴,⁸]hexadeca-1(15),4,6,8(16),
11,13-hexaene (traditional name:
[2.2]Metacyclophane)

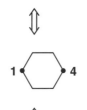

IUPAC: 1,4(1,3)-Dibenzenacyclohexaphane

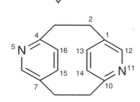

Chem. Abstr.:
5,11-Diazatricyclo[8.2.2.2⁴,⁷]hexadeca-1(12),
4,6,10,13,15-hexaene
IUPAC: 1(2,5),4(5,2)-Dipyridinacyclo-
hexaphane

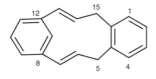

Chem. Abstr.:
5,15-Dihydro-12,8-metheno-8H-benzo-
cyclotridecene

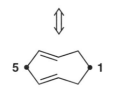

IUPAC: 1(1,2),5(1,3)-Dibenzenacyclo-
octaphane-3,6-diene

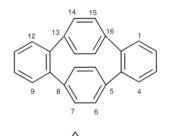

Chem. Abstr.:
5,8:13,16-Diethenodibenzo[*a,g*]cyclo-
dodecene

IUPAC:
1,3(1,2),2,4(1,4)-Tetrabenzenacyclo-
tetraphane

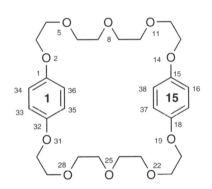

Chem. Abstr.:
2,5,8,11,14,19,22,25,28,31-Decaoxa-
tricyclo[30.2.2.2^{15,18}]octatriaconta-
1(34),15,17,32,35,37-hexaene
(triv.: Bis(*para*-phenylene)-[34]
crown-10)

IUPAC:
1,15(1,4)-Dibenzena-2,5,8,11,14,
16,19,22,25,28-decaoxacyclo-
octacontaphane

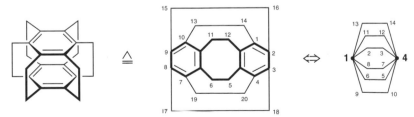

CA: 5,6,11,12-Tetrahydro-1,10:2,9:3,8:4,7-tetraethano-
       dibenzo[*a,e*]cyclooctene
IUPAC: 1,4(1,2,3,4,5,6)-Dibenzenapentacyclo[2.2.2.2.2.2]tetra-
        decaphane ("Superphane")

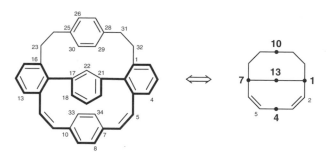

CA: 1,16-(Ethano[1,4]benzenoethano)-7,10-etheno-21,17-metheno-17*H*-dibenzo[*a,h*]cycloheptadecene

IUPAC: 1,7(1,2,3),4,10(1,4),13(1,3)-Pentabenzenabicyclo[5.5.1]tridecaphane-2,5-diene

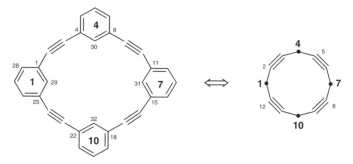

CA: Pentacyclo[23.3.1.1[4,8].1[11,15].1[18,22]]dotriaconta-1(29),4,6,8(30), 1,13,15(31)18,20,22(32),25,27-dodecaene-2,9,16,23-tetrayne

IUPAC: 1,4,7,10(1,3)-Tetrabenzenacyclododecaphane-2,5,8,11-tetrayne

That this new nomenclature concept can also be advantageously utilized with saturated, bridged, hetera-replaced etc. **cycla segments** is demonstrated by a supplemental example.

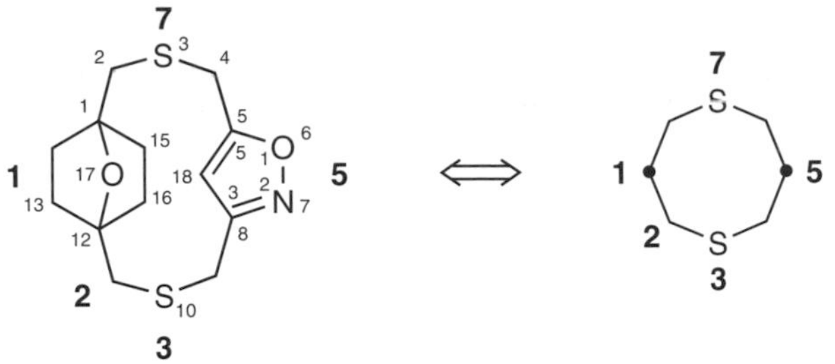

CA: 7-Aza-6,17-dioxa-3,10-dithiatetracyclo[10.2.2.1^{1,12}.1^{5,8}]dodeca-5(18),7-diene (small outer numbering)

IUPAC: 1^7-Oxa-3,7-dithia-1(1,4)-bicyclo[2.2.1]heptana-5(3,5)-1,2-oxazo lacyclooctaphane (fat numbering)

1.3.2
Other Phanes

Initially, the phane concept was intended solely for supercyclic systems, namely, **cyclophanes.** During elaboration of the new all-encompassing cyclophane nomenclature it became more and more obvious that its basic principles apply equally well to extended linear assemblies of chain and ring segments. Such **superchains** can then be treated exactly like **cyclophanes,** except that their names end with the morpheme ... **phane** after the numerical term denoting the length of the superchain.

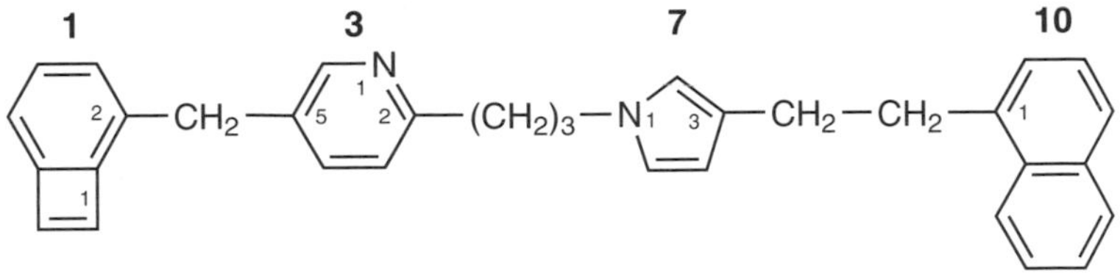

1(2)-Cyclobutabenzena-3(5,2)-pyridina-7(1,3)-pyrrola-10(1)-naphthalenadecaphane

A further extension of this system consists in the inclusion of spiro components and appears particularly advantageous in the case of multilayered polyspiro systems of the following type:

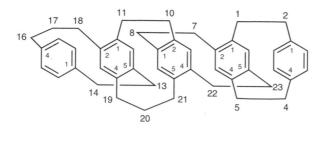

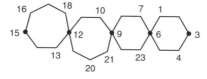

3.15(1,4),6(1,4,2,5),9(1,2,5,4),12(1,5,2,4)-pentabenzenatrispiro[5.2.2.6^{12}.3^9.2^6]triocosaphane

To close this chapter and in anticipation of subsequent discussions it should be kept in mind that the new assembly names obtained along the lines of **phane nomenclature** can now be used as parent names for the multifarious ways of substitution by functional groups in the same way as the more conventional parent names of old.

2 Substituted Systems

2.1
Preliminary General Remarks

The notion of a "substituent", which has already been repeatedly used in the discussions of substituent groups derived from parent structures, will obviously occupy a central position in the following sections and therefore requires a broadened definition.

> **The designation "substituent" pertains to any atom and any combination of atoms (functional group) that replaces a hydrogen atom of a parent structure.**

A broader specification of the notion "substituent" is obtained by subsuming all substituting groups other than hydrocarbon or hetero-cyclic/-catenic groups under the term **characteristic group.** These groups then determine the (**functional**) class; a **class name** must not necessarily reappear in the name of the individual members of the compound class in question, e.g.: –COOH carboxylic acid, –NO₂ nitro compound, –NH₂ amine or amino compound, –OH alcohol, etc. The trivial fact that a parent structure can simultaneously be substituted by several characteristic groups necessitates a **seniority order for substituents.** The most senior group then always defines the nominal compound class and is, whenever possible, expressed as a **suffix** to the parent name; other characteristic/functional groups present are abstracted in the form of appropriate **prefixes**, in alphabetical order and in compliance with the principle of lowest locants.

Multiplying prefixes are used in the same way as with parent structures: the series **di, tri, tetra …** for sets of identical substituents, the series **bis, tris, tetrakis …** for identically substituted identical substituents (or if linguistically better suited), and the series **bi, ter, quater …** for identical partial components directly joined together.

While the nomenclature procedures for parent structures have been comprehensively unified, for substituted systems the situation is much less comfortable since 1) differing nomenclature systems are still used quite unsystematically side by side, 2) many exceptional rulings are to be accounted for, and 3) many trivial and semitrivial names and rulings are still retained in equally unsystematic ways.

Specific nomenclature problems of a more complex nature can therefore frequently only be solved with the help of the stringently codified manuals of the **IUPAC** and **Chem. Abstr. rules** (see p. 5). Here, however, emphasis is placed on a concise survey of the general principles behind these naming conventions.

2.2
Nomenclature Types for Substituted Systems

2.2.1
Substitutive Nomenclature

According to this preferred type of nomenclature appropriately tailored morphems for substituent and characteristic groups are attached as prefixes or suffixes to the stem (parent) name. The substituents specified in Table 6, however, are exclusively expressed as prefixes. In the same way, substituting hydrocarbon and heterocyclic parent structures are prefixed in the form of their group names.

All other characteristic groups can be named both as prefixes and/or as suffixes. The **most senior characteristic group** is expressed as **suffix** and

Table 6. Characteristic groups expressed as prefixes only

Characteristic group	Prefix (Class name)
$-N_3$	Azido…(Azide)
$-Br$	Bromo…
$-Cl$	Chloro…
$-ClO$	Chlorosyl…
$-ClO_2$	Chloryl…
$-OCN$	Cyanato…Cyanate
$=N_2$	Diazo…
$-I(OH)_2$	Dihydroxyiodo … Systematic name: Dihydroxy-λ^3-iodanyl…
$-IX_2$	Di…iodo…(X is any acid group: $-Cl$, $-OOCCH_3$ etc.) system.: Di…λ^3-iodanyl
	Epithio… (traditional)

Table 6 (continued)

Characteristic group	Prefix (Class name)
	Epoxy... (traditional)
–F	Fluoro...
–OOH	Hydroperoxy...
–NHOH	Hydroxyamino... (Hydroxylamine)
–NHNH$_2$	Hydrazino... systematic name: Diazanyl...
–NCO	Isocyanato...
–NC	Isocyano... (Isocyanide, Isonitrile)
–NCS	Isothiocyanato...
–I	Iodo...
–IO	Iodosyl... (formerly Iodoso...)
–IO$_2$	Iodyl...
–NO$_2$	Nitro...
=N(O)OH	aci-Nitro... systematic name: Hydroxynitroryl...
–NO	Nitroso...
–ClO$_3$	Perchloryl...
–SCN	Thiocyanato...
–OOR	...yldioxy... (Peroxide)
–OR, –SR etc.	...yloxy..., ...ylthio... (Ether, Thioether etc.) system.: ...ylsulfanyl etc.

thus determines the compound type or class; the remaining groups are then denoted as prefixes in front of the stem name. The most senior group is identified with the aid of Table 7, which lists the most important compound classes in descending order of priority. Since this table is (regrettably) still quite fragmentary, the much more comprehensive seniority list of the **Chem. Abstr. Index Guide** (see literature on p. 5) may be consulted if necessary.

The manner in which characteristic groups pertaining to the compound classes listed in Table 7 are to be expressed in the form of prefixes and suffixes is shown in Table 8.

Whereas for most compound classes the exemplifications in Table 8 are directly applicable without any difficulty, the two alternative naming methods for aliphatic carboxylic acids and their derivatives as well as for nitriles and aldehydes require more detailed illustration.

Table 7. Seniority order of the most prominent compound classes

IUPAC	Chem. Abstr.
1. Radicals	1. Radicals
2. Anions	2. Cations
3. Cations	3. Neutral coordination compounds
4. Zwitterions	4. Anions

5. Carboxylic acids in the order: –COOH, –$\overset{O}{\text{C}}$OOH, then sulfur and selenium analogues, then sulf(onic, inic)-, phosph(onic, inic)-, ars(onic; inic)-acids
6. Acid derivatives in the order: anhydrides, esters, halides, amides, hydrazides, imides, amidines etc.
7. Nitriles (cyanides), isocyanides etc.
8. Aldehydes, S-, Se- and Te analogues
9. Ketones, S-, Se- and Te analogues
10. Alcohols, phenols, S-, Se, and Te analogues
11. Hydroperoxides, thiohydroperoxides etc.
12. Amines, imines, hydrazines, phosphanes etc.
13. Ethers, S-, Se-, and Te analogues
14. Peroxides, disulfides, etc.

To name simple aliphatic monoacids (and their derivatives) the carbon atom of the carboxylic group is included in the parent system, which is expressed by changing the hydrocarbon **...ane** ending to **...anoic acid.**

$$\overset{21}{\text{H}_3\text{C}}-(\text{CH}_2)_{18}-\overset{2}{\text{CH}_2}-\overset{1}{\text{COOH}} \quad \text{Henicosanoic acid}$$

In the case of multiple and cyclic carboxylic acids (and their derivatives) the acid group is treated as a unit corresponding to the suffix modes **...carboxylic acid, ...carboxamide, ...carbonyl chloride,** etc.

HO$_2$C
　　　＼
　　　　HC−CH=CH−CO$_2$H
　　　／
HO$_2$C

Prop-2-ene-1,1,3-
tricarboxylic acid

COCl
｜
N

H   COCl

Pyridine-1,4(4*H*)-di
(carbonyl chloride)

$$\overset{11}{\text{H}_3\text{C}}-\overset{10}{\text{C}}-\overset{9}{\text{CH}}-\text{CH}_2-\overset{7}{\text{CH}}-(\text{CH}_2)_3-\overset{3}{\text{C}}-\overset{2}{\text{CH}}-\overset{1}{\text{CH}_2}-\text{CONH}_2$$

(with substituents: CONH$_2$ at top of C-3 and C-2; S (double bond), CONH$_2$, CONH$_2$, BrS, CONH$_2$ below)

3-Bromosulfanyl-10-thioxoundecane-1,2,3,7,9-pentacarboxamide

Table 8. Prefixes and suffixes for the most important characteristic groups in substitutive nomenclature

Compound class	Characteristic group[a]	Use as prefix	Use as suffix
Cations	$\overset{\oplus}{-}OR_2$, $-NR_3^{\oplus}$, $-BrR^{\oplus}$ $-N{\equiv}N^{\oplus}$	...onio... Diazonio...	...onium... -diazonium...
Carboxylic acids	–COOH –COOH	Carboxy... –	-carboxylic acid -oic acid
Peroxycarboxylic acids	–C(O)OOH –C(O)OOH	? –	-peroxycarboxylic acid Peroxy...oic acid
(Di)thiocarboxylic acids	–COSH, –CSSH –COSH, –CSSH	(Di)thiocarboxy... –	-carbo(di)thioic S-acid -(di)thioic S-acid
Sulf$\begin{Bmatrix}on\\in\\en\end{Bmatrix}$ic acid	–SO₃H –SO₂H –SOH	Sulf$\begin{Bmatrix}o\\ino\\eno\end{Bmatrix}$...	-sulf$\begin{Bmatrix}on\\in\\en\end{Bmatrix}$-ic acid
Carboxylic acid salts	–COOM –COOM	M-carboxylato... –	M-...carboxylate M-...oate
Sulf$\begin{Bmatrix}on\\in\\en\end{Bmatrix}$ic acid salts	–SO₃M –SO₂M –SOM	M-sulf$\begin{Bmatrix}on\\in\\en\end{Bmatrix}$ato...	M-...sulf$\begin{Bmatrix}on\\in\\en\end{Bmatrix}$ate
Carboxylic acid anhydride	$-\underset{\parallel}{\overset{}{C}}-O-\underset{}{\overset{}{C}}-$ O O	–	...acid...acid anhydride
Carboxylic acid esters	–COOR –COOR	...yloxycarbonyl... –	...yl...carboxylate ...yl...oate

Table 8 (continued)

Compound class	Characteristic group[2]	Use as prefix	Use as suffix
Sulf $\begin{Bmatrix} on \\ in \\ en \end{Bmatrix}$ ic acid esters	$-SO_3R$ $-SO_2R$ $-SOR$	...yloxysulf $\begin{Bmatrix} on \\ in \\ en \end{Bmatrix}$ yl...	...yl...sulf $\begin{Bmatrix} on \\ in \\ en \end{Bmatrix}$ ate
Lactones	 	–	...olide, -carbolactone
Sultones		–	...sultone
Carboxylic acid halides	$-COX$ $-COX$	Halogenocarbonyl... –	-carbonyl halide -oyl halide[b]
Sulf $\begin{Bmatrix} on \\ in \\ en \end{Bmatrix}$ ic acid halides	$-SO_2X$ $-SOX$ $-SX$	X-sulf $\begin{Bmatrix} on \\ in \\ en \end{Bmatrix}$ yl...	-sulf $\begin{Bmatrix} on \\ in \\ en \end{Bmatrix}$ yl halide
Carboxylic acid amides	$-CONH_2$ $-CONH_2$	Aminocarbonyl... (trad. Carbamoyl) –	-carboxamide -amide
Carbonimidic acides	$-C(NH)OH$ $-C(NH)OH$	Hydroxy(imino)methyl... –	-carboximidic acid imidic acid
Lactams		–	-lactame

Class	Structure	Prefix	Suffix / Name
Lactimes	 (ring: R, N=, COH)	—	-lactime
Carboxylic acid imides	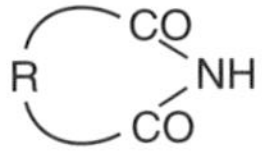 (ring: R, CO, NH, CO)	—	-dicarboximide
	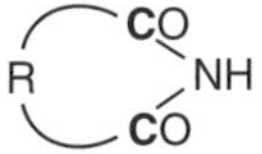 (ring: R, CO, NH, CO)	—	...imide (for trivial names)
Carboxylic acid hydrazides	–CONHNH₂ –CONHNH₂	Hydrazinocarbonyl —	-carbohydrazide ...ohydrazide
Hydrazonic acids	–C(NHNH₂)OH –C(NHNH₂)OH	Hydrazinohydroxymethyl... —	-carbohydrazonic acid ...hydrazonic acid
Hydroxamic acids	–CONHOH –CONHOH	(Hydroxamino)carbonyl... —	-carbohydroxamic acid ...hydroxamic acid
Hydroximic acids	–C(NOH)OH –C(NOH)OH	Hydroximinohydroxymethyl... —	-carbohydroximic acid ...hydroximic acid
Amidines	–C(NH)NH₂ –C(NH)NH₂	Carbamimidoyl[c]... [Amino(imino)methyl...]	-carboximidamide[c] ...imidamide[c]
Amidoximes	–C(NOH)NH₂ –C(NOH)NH₂	Amino(hydroximino)methyl... —	-carboxamidoxime ...amidoxime
Sulf{on/in/en}ic acid amides	–SO₂NH₂ –SONH₂ –SNH₂	Sulf{amoyl/inamoyl/enamoyl}[d]...	...sulf{on/in}amides ...sulf{en}...
Sulf{on/in}imidic acids	–S(O)(NH)OH –S(NH)OH	Hydroxy(imino)sulfinyl... Hydroxy(imino)-λ⁴-sulfanyl...	-sulf{on/in}imidic acid

Table 8 (continued)

Compound class	Characteristic group[2]	Use as prefix	Use as suffix
Sultames	 R⟨NH, –, SO₂⟩	–	...sultame
Sulf⟨on/in/en⟩ acid hydrazides	$-SO_2NHNH_2$ $-SONHNH_2$ $-SNHNH_2$	Hydrazino⟨sulfonyl/sulfinyl/sulfanyl⟩...	-sulf⟨ono/ino/eno⟩ hydrazide
Nitriles	$-C\equiv N$ $-C\equiv N$	Cyano... –	-carbonitrile ...nitrile
Aldehydes	$-CHO$ $-CHO$	Formyl... Oxo...	-carbaldehyde ...al
Thioaldehydes	$-CHS$ $-CHS$	Thioformyl... Thioxo...	-carbothialdehyde thial
Ketones	$>C=O$	Oxo...	...one
Thioketones	$>C=S$	Thioxo...	...thione
Acetals, Ketals	$>C(OR)_2$	Di-(...yloxy)...	al⟩di...yl⟨-acetal ...one⟩ ⟨-ketal
Oximes	$>C=NOH$	Hydroximino...	al⟩oxime ...one⟩
Hydrazones	$>C=NNH_2$	Hydrazono...	al⟩hydrazone ...one⟩
Azine	$>C=N-N=C<$	Azinodi...	al⟩azine ...one⟩
Semicarbazones	$>C=NNHCONH_2$	Semicarbazono...	al⟩semicarbazone ...one⟩

Alcohols, Phenols	–OH	Hydroxy...	...ol
Thiols	–SH	Sulfanyl... (Mercapto...)	...thiol
Alcohol } Phenol } ates	–OM	M-oxido...	M...olate
Thioalcohol } Thiophenol } ates	–SM	M-sulfido...	M...thiolate
Amines	–NH$_2$	Amino...	...amine[b]
Imines	=NH	Imino...	...imine[b]

[a] C atoms in bold type are included in the stem name.
[b] See also radicofunctional nomenclature
[c] Formerly: Amidino-, -carboxamidine- and -amidine, respectively.
[d] Systematically: amino(sulfonyl, sulfinyl, sulfanyl)...

This table can easily be logically extended. Thus, for example, the naming procedures for sulf(onic, inic, enic) acids and their derivatives are directly transferable to thio analogues. Several of the characteristic groups presented here can be further substituted in their –NH$_2$ and –OH functions. In the respective names this is accounted for by prefixing the corresponding group name including its locant, e.g.: N-methyl-hexanamide.

Corresponding nitriles and aldehydes are treated in the same manner:

$$H_3C-(CH_2)_5-CH_2- \begin{cases} C\equiv N \\ C\underset{O}{\overset{H}{\diagup}} \end{cases} \quad \text{Octane} \begin{cases} \text{nitrile} \\ \text{al} \end{cases}$$

$$\begin{array}{c} CH_2C\equiv N \\ | \\ HO-C-C\equiv N \\ | \\ CH_2C\equiv N \end{array}$$

2-Hydroxypropane-
1,2,3-tricarbonitrile

Biphenyl-2,2'-
dicarbaldehyde

It should be emphasized once again that the rules of **substitutive nomenclature are to be preferred over all other naming methods** because they are indeed most generally and most extensively applicable. Nevertheless, for a number of not always clearly delineated types of compounds several other nomenclature systems – to be outlined in the following sections – are still retained; it is hoped, though, that these will gradually be abolished in favour of substitutive nomenclature.

2.2.2
Functional Class Names (Formerly: Radicofunctional Nomenclature)

This naming method is used only for a limited number of compound classes which are listed in Table 9 in descending order of priority. In contrast to substitutive nomenclature, **parent systems** are named here **in the guise** of their **"radicals"**, i.e. **substituent groups,** to which are added, separated by a space, the **anionized names** of the compound class in question.

Some examples may illustrate the principles of this nomenclature type:

$$\begin{array}{c} H_2C \\ \diagdown \\ \diagup \quad CH-C\underset{Cl}{\overset{O}{\diagup}} \\ H_2C \end{array} \quad \text{Cyclopropanecarbonyl chloride}$$

H_2CCl_2 Methylene dichloride
(mostly: Methylene chloride)

H_3C-CN Methyl cyanide

H_3C-OCH_3 Dimethyl ether

$(C_6H_5)_3N$ Triphenylamine

$H_3C-CH_2-CH_2-CH_2-CH_2-COF$
Hexanoyl fluoride

$H_3C-CO-C_2H_5$ Ethyl methyl ketone

$(C_6H_5)_2SO$ Diphenyl sulfoxide

$C_6H_5CH_2-N=C=O$ Benzyl isocyanate

Table 9. Compound names as used in radicofunctional nomenclature (Functional Class Names)

Characteristic group	Radicofunctional name
X in acid derivatives RCO–X, RSO$_2$–X usw.	Anionic name of X in the order: fluoride, chloride, bromide, iodide, cyanide, azide, etc., then S and Se analogues
–C≡N:, –N≡C: (⊕ ⊖)	cyanide, isocyanide
–O–C≡N:; –N=C=O	cyanate, isocyanate
–O–N≡C: (⊕ ⊖)	fulminate
–S–C≡N:, –N=C=S	thiocyanate, isothiocyanate etc.
⟩C=O, ⟩C–S	ketone, thioketone etc.
⟩C=C=O	ketene
–OH, –SH	alcohol, hydrosulfide
–O–OH, –S–S$_x$–SH	hydroperoxide, hydropolysulfide
–O–, –O–O–	ether or oxide, peroxide
–S–, –S–S$_x$–S–	sulfide, polysulfide
⟩S=O, ⟩SO$_2$	sulfoxide, sulfone
⟩S=NH, ⟩S(O)NH	sulfimide, sulfoximide
⟩Se, ⟩SeO, ⟩SeO$_2$	selenide, selenoxide, selenone
–F, –Cl, –Br, –I, –N$_3$	fluoride, chloride, bromide, iodide, azide
RNH$_2$, RR'NH, RR'R''N	amine
⟩C=N–C⟨	amine (azomethine, Schiff's bases)
⟩C(OOCR)$_2$	-di...oate(acylals)

2.2.3
Additive Nomenclature

As already apparent from its name, this nomenclature type is based on the addition of atoms or groups of atoms to a parent structure and is restricted to very few special cases. The most important and at any rate indispensable application of this method is found in the naming of **hydrogenated fused polycycles** where hydrogenation is indicated by appropriate prefixes.

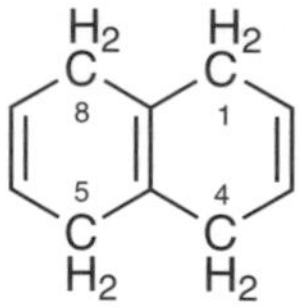

1,4,5,8-Tetrahydronaphthalene

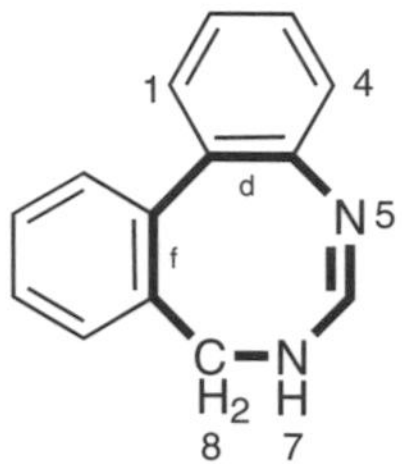

7,8-Dihydrodibenzo[d,f][1,3]diazocine

In contrast, **epoxides, ozonides,** and certain **halogen derivatives** are characterized by **anionized name forms** of the added atoms (or groups of atoms) placed behind the stem terms. It should be noted, though, that retention of this method is not recommended when other types of nomenclature (substitutive or heterocycle nomenclature) permit unambiguous simpler names.

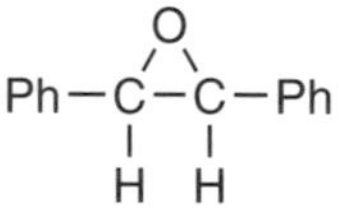

Stilbene oxide 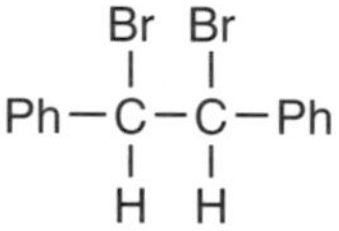   Stilbene dibromide

The designation **ozonide** should be used only if no exact structure of the compound in question is known; otherwise a heterocyclic name is to be preferred.

$C_3H_6O_3$ Propene ozonide   but:

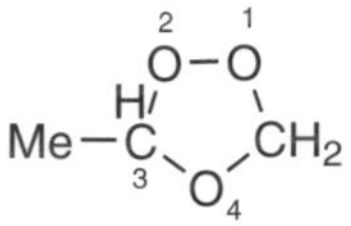

3-Methyl-1,2,4-trioxolane

For compounds of the amine oxide and nitrile oxide type and analogous systems with **oxidized heterocenters** (above all heterocyclic systems) additive nomenclature alone offers reasonable names.

$(C_4H_9)_3\overset{\oplus}{N}-\overset{..}{\underset{..}{O}}:\ominus$ Tributylamine oxide

$C_{12}H_{25}-C\equiv\overset{\oplus}{N}-\overset{..}{\underset{..}{O}}:\ominus$ Tridecanenitrile oxide

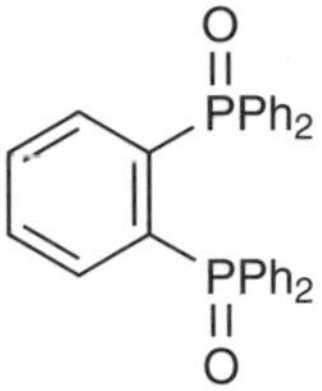

1,2-Phenylenebis(diphe-
nylphosphane oxide)
(Benzene-1,2-diyl) bis(di-
phenylphosphane oxide)

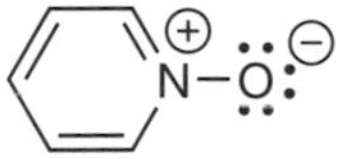

Pyridine oxide    Thietane-1,1-dioxide

Nitrones too are named as amine oxides:

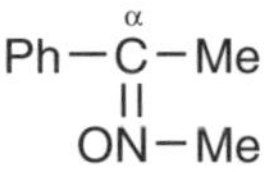

$$Ph-\overset{\alpha}{\underset{\underset{ON-Me}{||}}{C}}-Me$$ N-(α-Methylbenzylidene)methanamine N-oxide

Another variant of additive nomenclature, where the increase of chain or ring links by one CH_2 unit is indicated by the prefix **homo**, finds application mainly for steroid names (see Table 22, p. 200) but also for certain trivially named compound types.

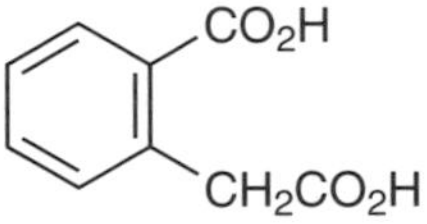

CO$_2$H

CH$_2$CO$_2$H

Homophthalic acid    Tris(homo)benzene

Likewise restricted mainly to natural products (but with some foreseeable potential also for simpler compound types) is the use of **seco** for indicating ring cleavage with concomitant **addition of two hydrogen atoms** to the carbon atoms involved.

Yohimbane (fundamental)
parent structure)

2,3-Secoyohimbane

2.2.4
Subtractive Nomenclature

This nomenclature type uses **substractive prefixes** to indicate removal of atoms or groups of atoms from a trivially or systematically named parent structure. While additive nomenclature is used above all for hydrogenated cyclic systems, subtractive nomenclature focuses primarily on the opposite phenomenon, i. e. **introduction of unsaturation.** This has already been dealt with extensively in the discussions relating to hydrocarbon systems, where the terminal syllables **... ene** and **... yne** signify loss of two and four hydrogen atoms with concomitant formation of a double and triple bond, respectively. A more explicit symbolization of dehydrogenation in the guise of a **subtractive prefix** is particularly recommendable for certain natural products of the steroid and carbohydrate series as well as for **dehydro**arenes (less precisely also named **arynes**) and **dehydro**annulenes.

1,2-Didehydrobenzene
(trivially = Benzyne,
Dehydrobenzene)

2,6-Didehydropyridine
(Pyridine-2,6-diyl)

1,2,7,8-Tetradehydro[12]annulene

Also restricted mainly to complex and/or trivially named natural products (e.g.: **alkaloids**) remains a method for indicating replacement of *N*-methyl groups by hydrogen with the prefix **de**...

$$\ge N-CH_3 \longrightarrow \ge N-H \qquad De-N\text{-methyl}...$$

Similar limitations apply for the specification of an analogous replacement of hydroxy groups by hydrogen, as occurs frequently in carbohydrate chemistry.

$$-CH_2OH \longrightarrow -CH_3, -CHOH- \longrightarrow -CH_2- \text{ etc.} \qquad Deoxy...$$

Several compound classes obtainable by dehydration of appropriate precursors have **subtractive class names:** anhydrides, lactones, sultams, etc. The prefix **anhydro** for indicating removal of water from two hydroxy groups is again mainly used for carbohydrates (see Chapter 5).

$$
\begin{array}{ll}
HC^x-OH & HC \\
\quad | y & \quad | \\
HO-CH & HO-CH \diagdown O \quad x,z\text{-Anhydro}... \\
\quad | & \quad | \diagup \\
HC_z-OH & HC
\end{array}
$$

Special conventions apply for the use of the prefix **nor**..., another manifestation of subtractive terminology. In the **terpene** series, **nor**-compounds are those which have undergone replacement of all methyl groups on a ring by hydrogen atoms (see Appendix, Table 21). For **steroids** (Appendix, Table 22) **nor**... means first of all **loss of a CH₂ group** from a chain and, in a second sense, **ring contraction** with the expulsion of **one CH₂ unit**, i.e. the opposite of **homo**....

Out of the specific necessities of steroid nomenclature another highly useful subtractive descriptor came into being, namely, **cyclo**..., for designating an **additional direkt link** between any non vicinal C atoms of a parent structure with concomitant **loss of two H atoms.**

Estrane (fundamental parent structure)

7,11α-Cyclo-7,8-seco-17a-homo-5α-estrane

It is tempting to envisage use of the **cyclo procedure** also for very simple parent systems, above all when special relationships could thus be expressed by the name itself, e. g.:

1,4-Cyclobenzene
(Bicyclo[2.2.0]hexa-1,3,5-triene)

3,5-Cyclopyridine
(Cyclopropa[c]pyrrole)

Another very specific subtractive naming mode has been developed for cyclic compounds containing (if only formal) contiguous double bonds – **cyclic cumulenes** – whose treatment with conventional nomenclatural means would often lead to rather cumbersome combinations of "ene", "dehydro", and "indicated H" notations.

Substantial relief is provided here by the so-called **delta convention** whose essence consists in identifying any skeletal atom at which two (or more) double bonds (e. g.: $m = 2,3$) converge with a δ^m-**symbol** placed immediately behind the corresponding locant and λ^n-symbol, if present. (Pure Appl. Chem. **1988**, *60*, 1395).

$5\delta^2$-Dibenzo[*a,d*]cycloheptene
instead of:
4a,5-Didehydro-4a*H*-dibenzo[*a,d*]cycloheptene

8,8a-Dihydro-$6\delta^2$-[1,3]diphosphinino]1,2-*d*][1,2,4]oxazaphosphole
instead of:
5,6-Didehydro-8,8a-dihydro-5*H*-[1,3]diphosphinino...

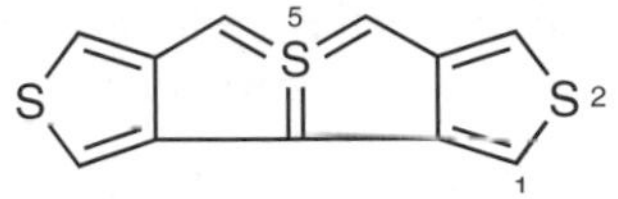

$5\lambda^6\delta^3$-Thieno[3,4-*c*]thieno[3',4':3,4]thieno[1,2-*a*]thiophene
instead of:
1,5,3,5-Tetradehydro-1*H*,3*H*-5λ^6-thieno

derived from:

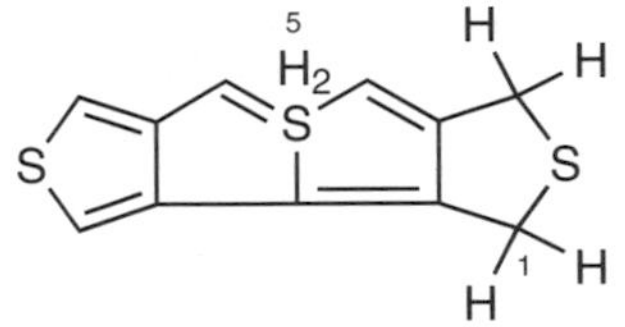

Once again, this method might prove advisable also for very simple systems, above all when special generic relationships warrant pointing out.

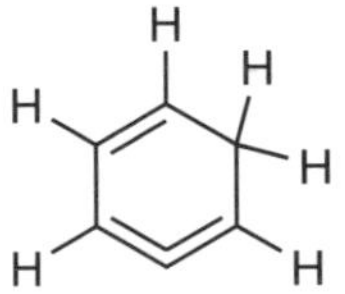

Trivial name: Isobenzene
Systematic name: Cyclohexa-1,2,4-triene
δ-Convention: $6H$-2δ^2-Benzene

2.2.5
Conjunctive Nomenclature

In principle this nomenclature types offers nothing that could not be settled equally well and, above all, more uniformly by substitutive nomenclature. Since, however, it makes allowance for the generation of new index names encompassing larger molecular units it has evolved into an irrevocable key element of **Chem. Abstr. registry nomenclature.** Thus, for example, **phenylacetic acid** would have to be looked for under the very wide index entry **acetic acid** whereas the autonomous conjunctive index entry **benzeneacetic acid** permits a much more efficient search. Acyclic **carboxylic acids, nitriles, aldehydes, alcohols,** and **amines** in particular, bearing a terminal cyclic substituent, are named conjunctively by combining the **unchanged stem name** of the **cycle** with the **unchanged name** of the **acyclic component.**

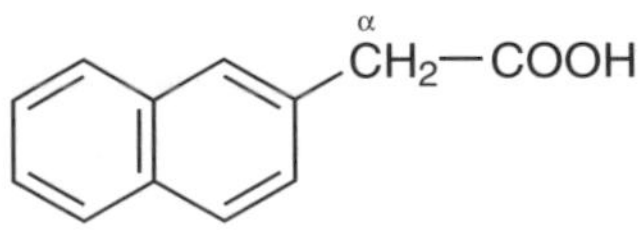

conjunctively: Naphthalene-2-acetic acid
substitutively: 2-Naphthylacetic acid

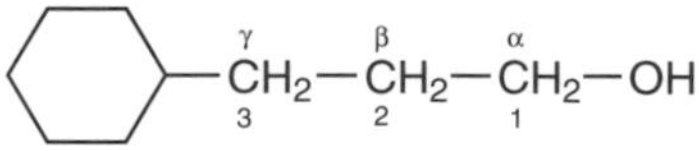

conjunctively: Cyclohexane-
                propanol

substitutively: 3-Cyclohexyl-
                propanol

but:

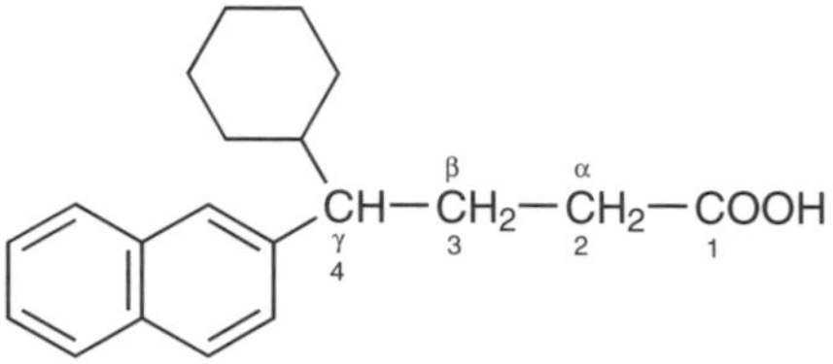

conjunctively: γ-Cyclohexyl-
                naphthalene-2-
                butanoic acid

substitutively: 4-Cyclohexyl-4-
                (2-naphthyl)-
                butanoic acid

Especially for systems with several identical substituents of senior parent structure character the conjunctive method offers manifest advantages.

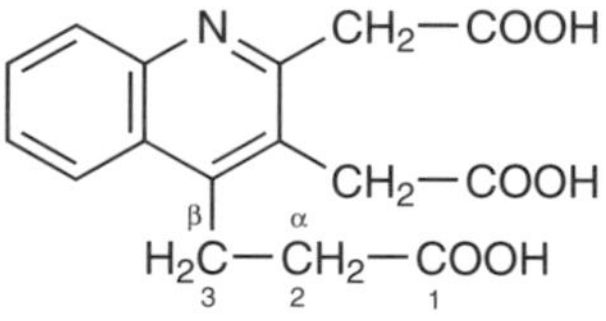

conjunctively: 4-(2-Carboxyethyl)iso-
                quinoline-2,3-diacetic acid

substitutively: 3-[2,3-Bis(carboxymethyl)
                isoquinoline-4-yl]pro-
                panoic acid

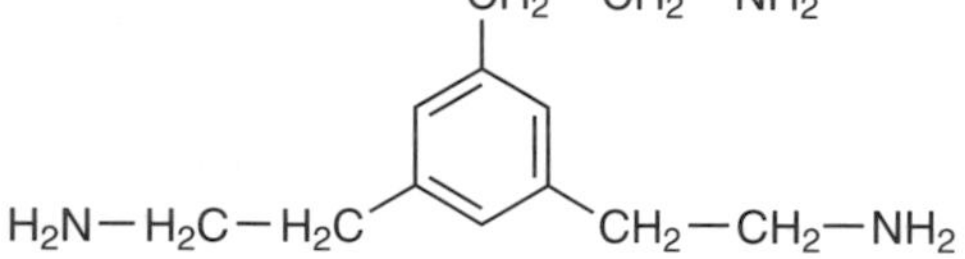

conjunctively: Benzene-1,3,5-triethanamine

substitutively: 2,2',2''-(Benzene-1,3,5-triyl)trisethanamine

2.2.6
Naming of Substituted Assemblies of Identical Units

As has already been conveyed in the discussions relating to parent structures, specific naming rules (on whose necessity opinions are divided) are to be obeyed for compounds assembled from identical cyclic components. The same holds for assemblies of identically substituted subunits, of which two types exist: those linked directly by single or double bonds and those which are substitutively connected through a bi- or polyvalent group.

2.2.6.1
Components with Direct Linkage

Here, first of all, the underlying unsubstituted parent components are joined together as described earlier (p. 38, 64). To take account of the substituents, in a second step these assemblies are subjected to the rules of substitutive nomenclature just like normal parent structures (lowest possible locants for positions of attachment!).

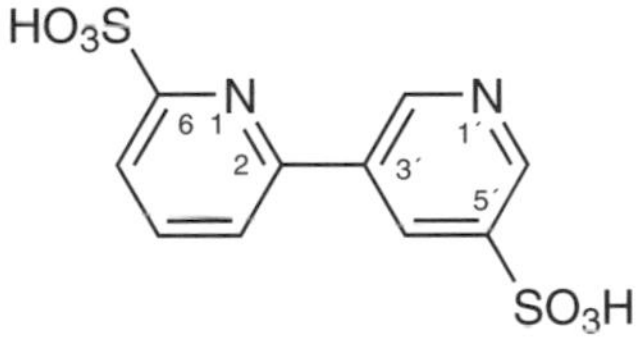

2,3'-Bipyridine- or: 2,3'-Bipyridyl- } 5',6-disulfoic acid

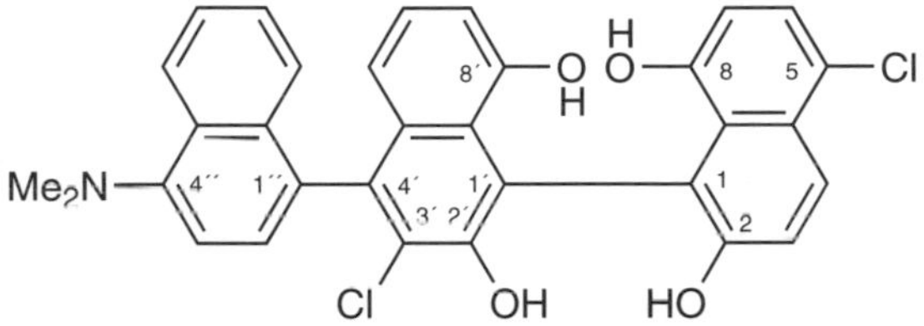

3',5-Dichloro-4"-(dimethylamino)-1,1':4',1"-ternaphthalene-2,2',8,8'-tetrol

The original numbering is, however, preserved when trivial names can be used for substituted partial components.

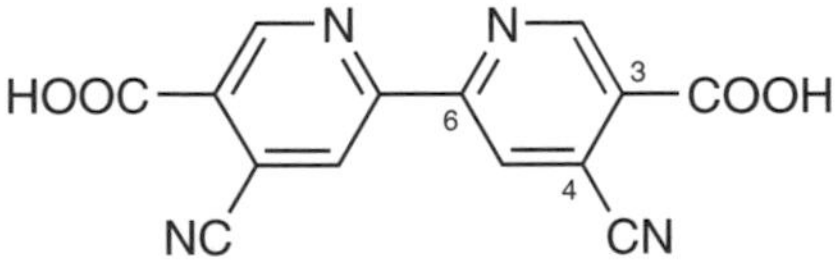

4,4'-Dicyano-6,6'-binicotinic acid

Formerly this naming procedure was occasionally also used for systems assembled from identical acyclic subunits.

$H_3C-CO-CO-CH_3$ HOOC–NH–NH–COOH $C_6H_5CH_2-CH_2C_6H_5$
Biacetyl   Bicarbamic acid   Bibenzyl

2.2.6.2
Identical Components Bound to Di- or Polyvalent Groups

Such compounds are named by first marking the positions of attachment of the parent components with the central group, then the name of that group followed by the multipliers **di...**, **tri...** (or **bis...**, **tris...**) etc., and finally the name of the parent component including its characteristic group(s). Additional substituents are accounted for by prefixes, with the following priorities of numbering being observed:

1. characteristic groups (suffixes)
2. positions of attachment
3. prefix substituents

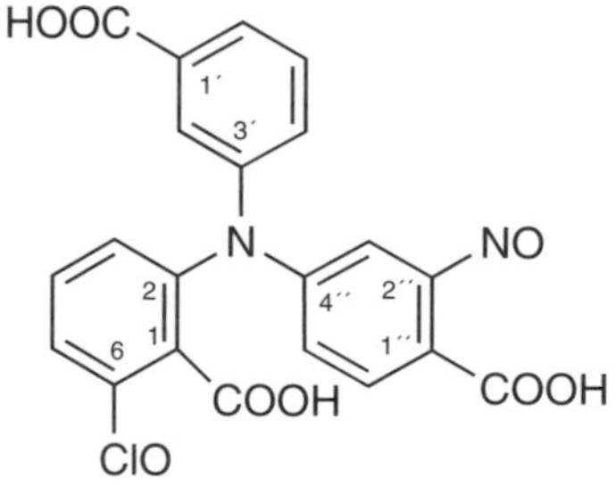

6-Chlorosyl-2″-nitroso-2,3′,4″-
nitrilotribenzoic acid

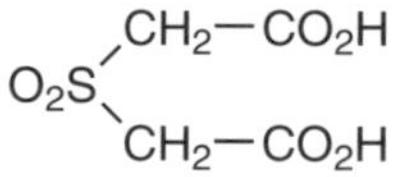

Sulfonyldiacetic acid

If the connecting groups are themselves assemblies of several parts, this concept is logically extended.

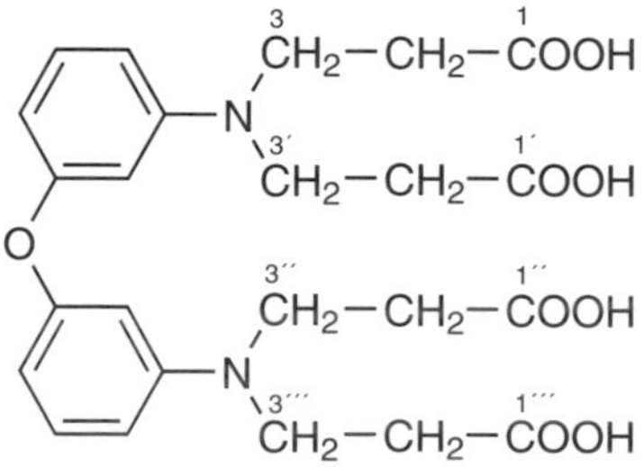

3,3′,3″,3‴-[Oxybis(1,3-phenylenenitrilo)]tetrapropanoic acid

The general applicability of such complex connecting groups as bivalent or polyvalent substituent groups is documented by some additional examples:

1,8-Naphthylene-dioxy... better: Naphthalene-1,8-diylbis(oxy)	Dithiodi-2,1-phenylene... new: Disulfanyldi-2,1-phenylene	Phosphanetriyltri-3,1-phenylenetris(methylene)... or: Phosphanylidynetri-3,1-...

Table 10, finally, presents a compilation of the most important bi- and polyvalent center groups suitable for this nomenclature type.

Table 10. Names for central connecting groups as used in the nomenclature for assemblies of identical units

Traditional	Formula	Systematic
Ethylene...	$-CH_2-CH_2-$	Ethane-1,2-diyl
Ethylenedioxy...	$-O-CH_2CH_2-O-$	Ethane-1,2-diylbis(oxy)
Azino...	$=N-N=$	Diazanebis(ylidene)
Azo...	$-N=N-$	Diazenediyl
Carbonimidoyl...	$-C(NH)-$	Iminomethylene
Carbonyl...	$-CO-$	
Carbonyldioxy...	$-O-CO-O-$	
Dioxy...	$-O-O-$	
Dithio...	$-S-S-$	Disulfanediyl
Hydrazo...	$-NH-NH-$	Diazane-1,2-diyl
Imino...	$-NH-$	
Methylene...	$-CH_2-$	
Methylenedioxy...	$-O-CH_2-O-$	
Naphthylenebisazo...	$-N_2C_{10}H_6N_2-$	...bis(diazenediyl)
Nitrilo...	$-N\!\!<$	
Oxy...	$-O-$	
Phenylene...	$-C_6H_4-$	
Phenylenebisazo...	$-N_2C_6H_4N_2-$	...bis(diazenediyl)
Polythio...	$-S_n-$	Polysulfanediyl
Seleninyl...	$-SeO-$	
Seleno...	$-Se-$	Selanediyl
Selenonyl...	$-SeO_2-$	
Sulfinyl...	$-SO-$	
Sulfonyl...	$-SO_2-$	

Table 10 (continued)

Traditional	Formula	Systematic
Sulfonyldioxy...	$-O-SO_2-O-$	
Tetramethylene...	$-CH_2(CH_2)_2CH_2-$	Butane-1,4-diyl
Thio...	$-S-$	Sulfanediyl
Thiocarbonyl...	$-CS-$	
Trimethylene...	$-CH_2CH_2CH_2-$	Propane-1,3-diyl
Trimethylenedioxy...	$-O(CH_2)_3O-$	Propane-1,3-diylbis(oxy)
Ureylene...	$-NH-CO-NH-$	

2.2.7
Naming of Radical and Ionic Species

In principle, all radicals and ions can nowadays be named in a uniform and totally systematic manner on the basis of the **operational suffixes** (and **prefixes derived therefrom**) compiled in Table 11. Obviously, standardization of the nomenclature for such derived species can be fully congruous only if the names of the underlying parent structures themselves are generated in a thoroughly systematic way, e.g.: **oxidane, dioxidane, azane, diazane,** etc. Hence, in the subsequent sections fully systematic names are always given as well as the conventional trivial/traditional designations.

Table 11. Operational affixes derived from parent hydrides by the indicated processes

	Suffix	H·	Prefix	Suffix	H⊕	Prefix	Suffix	H:⊖	Prefix
Addition	–		–	-ium[b]		-iumyl[b]	-uide		-uidyl
Subtraction	-yl[a]		-ylo	-ide[c]		-idyl[c]	-ylium		-yliumyl

[a] Subtraction of two and three H-atoms is expressed by: **-ylidene, -ylidyne,** with the exception of methylene ($:CH_2$), silylene ($:SiH_2$) etc., and aminylene ($:\dot{N}H$).

[b] For cations comprising core elements of groups 15–17 the traditional **-onium** suffixes and **-onio-** prefixes are still prefered.

[c] For anions of acids, alcohols, sulfides etc. the traditional suffixes **-ate, -oxide** and **-sulfide** and prefixes **-ato-, -ido-** are still retained.

2.2.7.1
Free Radicals

Electrically neutral groups containing unpaired electrons are designated as free radicals; their names are characterized by the ending **...yl**. The way in which names of parent structures are transformed into "radical" names in the substituent group sense has been indicated in the corresponding foregoing sections; the group names thus obtained can now directly be employed to name **factual free radicals**. A couple of further examples may illustrate this.

•CH_2–CH_2–$\overset{\bullet}{C}H$–CH_2–$\overset{\bullet}{C}H_2$ Pentane-1,3,5-triyl

•CH_2–CH_2–$\overset{\bullet\bullet}{C}H$ Propane-1-yl-3-ylidene (from an operational point of view propane-1,1,3-triyl would certainly be the better name!)

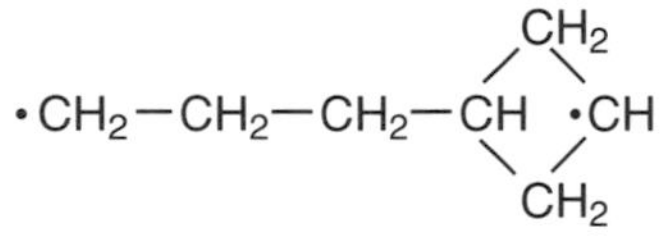

•CH_2—CH_2—CH_2—CH •CH 3-(3-Ylopropyl)cyclobutyl

(C_6H_5)Si: Diphenylsilylene   (here again diphenylsilanediyl would appear more appropriate)

Delocalized radicals can be named by way of their specific resonance forms or with traditional **summary group designations** devoid of any locants.

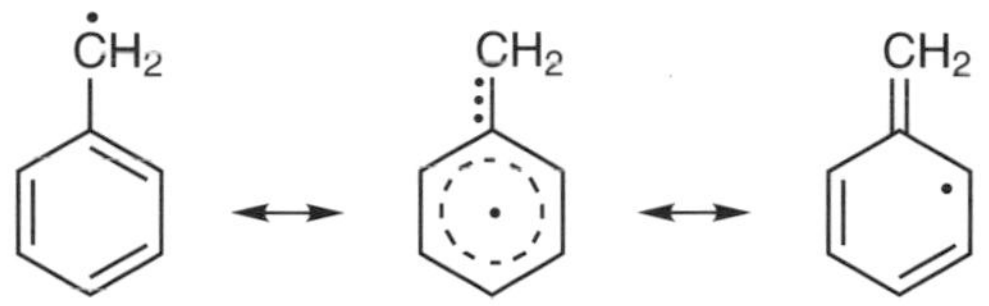

Phenylmethyl   5-Methylenecyclohexa-2,4-dien-1-yl

Benzylradical

Prop-2-en-1-yl   Allyl radical

The names of oxygen radicals which are mostly derived from acids, alcohols, etc. are traditionally assiged the ending ...**oxyl** or, systematically, the suffix ...**oxidanyl.** Radical names for other hetero groups can frequently be adopted straightforwardly from the corresponding substituent prefixes. For clarification they are usually supplemented by the descriptive term "**radical**".

$C_5H_{11}-\overset{..}{\underset{..}{O}}\cdot$ Pentyloxyl
Pentyloxy radical
(Pentyloxidanyl)

$C_6H_5-CO-\overset{..}{\underset{..}{O}}\cdot$ Benzoyloxyl
Benzoyloxy
radical
(Benzoyloxidanyl)

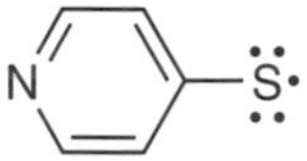

Pyridine-4-sulfenyl,
(Pyridine-4-ylsulfanyl)

$(H_3C)_3C-O-\overset{..}{\underset{..}{O}}\cdot$ *tert*-Butylperoxyl,
(*tert*-Butyldioxidanyl)

$H_3C-\overset{\cdot}{S}O_2$ Methanesulfonyl
(Methyldioxo-λ^6-sulfanyl)

Terephthaloyl
[1,4-Phenylene-bis(oxomethyl)]

Mono- and diradicals derived from amines are traditionally called **aminyls** and **aminylenes**, respectively.

$(C_6H_5)_2\overset{\cdot}{\underset{..}{N}}\colon$ Traditional name: Diphenylaminyl
Chem. Abstr.: *N*-Phenylbenzenaminyl
Systematic name: Diphenylazanyl

$H_3C-\overset{\cdot}{\underset{..}{N}}\cdot$ Methanaminylene, often also called methylnitrene
Systematic name: Methylazanylidene
(or methylazanediyl)

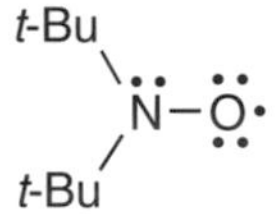

t-Bu
    $\overset{..}{\underset{..}{N}}-\overset{..}{\underset{..}{O}}\cdot$ Di-*tert*-butylaminyloxyl
t-Bu   Systematic name: Di-*tert*-butylazanyloxidanyl

$(C_6H_5)_2\overset{\cdot}{P}O$ Diphenylphosphinoyl
Systematic name: Oxodiphenyl-λ^5-phosphanyl

Radicals derived from hydrazine are traditionally named as **hydrazyls.**

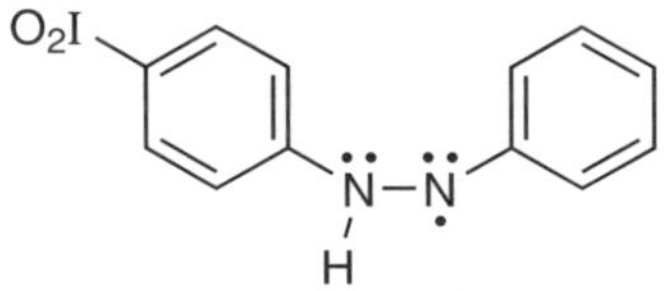

2-(4-Iodylphenyl)-1-phenylhydrazyl
Systematic name: ...diazanyl

2.2.7.2
Cations

In documenting the nomenclature rules for organic cations, four different naming modes can be defined; treatment of cations within the conventions of "a" nomenclature has already been noted on p. 52.

a) Cations derived from the ... **onium** prototypes compiled in Table 12.

Table 12. "Onium" prototypes in descending order of priority

Ion	Class name	Cation as prefix
$H_4N^\oplus$	Ammonium	Ammonio
$H_4P^\odot$	Phosphonium	Phosphonio
$H_4As^\oplus$	Arsonium	Arsonio
$H_4Sb^\oplus$	Stibonium	Stibonio
$H_4Bi^\oplus$	Bismuthonium	Bismuthonio
$H_3O^\oplus$	Oxonium	Oxonio
$H_3S^\oplus$	Sulfonium	Sulfonio
$H_3Se^\oplus$	Selenonium	Selenonio
$H_3Te^\oplus$	Telluronium	Telluronio
$H_2F^\oplus$	Fluoronium	Fluoronio
$H_2Cl^\oplus$	Chloronium	Chloronio
$H_2Br^\oplus$	Bromonium	Bromonio
$H_2I^\oplus$	Iodonium	Iodonio

$(C_2H_5)_2\overset{\oplus}{\underset{\bullet\bullet}{O}}H$ Diethyloxonium,

2,2'-Biphenylyleniodonium
(Biphenyl-2,2'-diyliodonium)

(4-Dimethylsulfoniophenyl)trimethyl-ammonium

b) Cations formed by fixation of a proton (hydron, see. p. 183) or another positive group onto a heteroatom of a trivially or systematically named compound are specified by the ending ...**ium** attached to that name (see also p. 63). If necessary such a proton is treated like an indicated hydrogen.

H$_3$C–$\overset{\oplus}{N}$H$_2$–NH$_2$ 1-Methylhydrazinium   system.: 1-Methyldiazanium

C$_6$H$_5$–CO–$\overset{\oplus}{N}$H$_3$ *NH*-Benzamidium   system.: Benzoylazanium

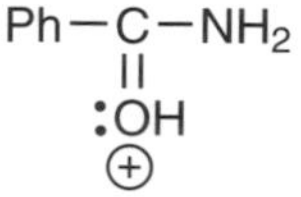

Ph–C–NH$_2$
$\parallel$
:OH
$\overset{(+)}{}$

$\overset{\oplus}{O}$*H*-Benzamidium,   systematic name: α-Amino-
benzylidenoxidanium

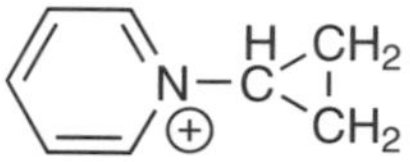

1-Cyclopropylpyridinium

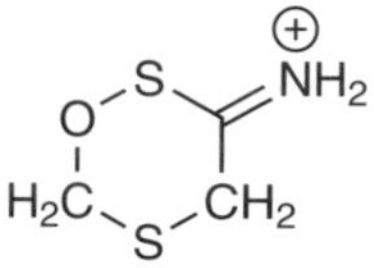

1,2,5-Oxadithian-3-iminium
(1,2,5-Oxadithian-3-ylidenazanium)

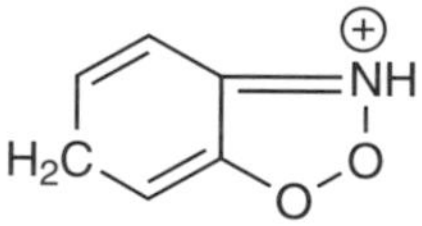

6*H*-1,2,3-Benzodioxazol-3-ium

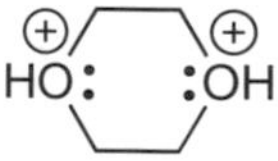

1,4-Dioxane-1,4-diium

c) Cations formed by unspecified addition of a hydron to an aromatic system are also named by attaching the ending ...**ium** to the name of the
parent. These so-called **arenium ions** correspond essentially to the π-
complexes of electrophilic aromatic substitution. (For specific addition
of hydrons to such systems see d.)

[C$_6$H$_7$]$^\oplus$ or 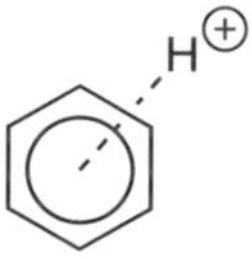   Benzenium   [C$_{14}$H$_{11}$]$^\oplus$ Anthracenium

The same procedure is applicable for depicting hydron addition to a saturated hydrocarbon as effected by superacids. The resulting **"nonclassical" carbocations** are subsumed under the class name **carbonium ions.**

$CH_5^\oplus$ Methanium $H_3C-CH_3^\oplus-CH_3$ =

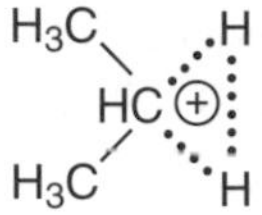

Propan-2-ium

Nonclassical cations derived from other saturated element hydrides can be named in like manner.

$BH_4^\oplus$ Boranium $NH_5^{2\oplus}$ Azanediium   or Azanebis(ium)

d) If cations are formally produced by removal of an electron from a radical position or by way of detachment of a hydride (protide, see p. 183) ion from a parent hydride, this is indicated by the composite name ending **...ylium**. Once again naming modes such as **...yl cation** have more descriptive character. The traditional class name for such species is **carbenium ions.**

$H_3C-CH_2^\oplus$ Ethylium
                Ethyl cation

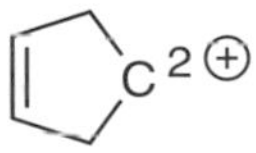

C $2\oplus$ Cyclopent-3-ene-1,1'-bis(ylium)
                              Cyclopent-3-enylidene dication

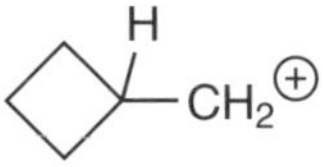

$CH_2^\oplus$ Cyclobutyl-
methylium
Cyclobutylmethyl
cation

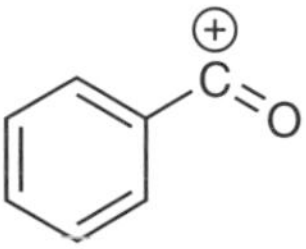

Benzoylium
Benzoyl cation
(Oxophenyl-
methylium)

$H_3C-SO_2^\oplus$ Methanesulfonylium; methanesulfonyl cation
                Fully systematic: Methyldi(oxo)-λ^6-sulfanylium

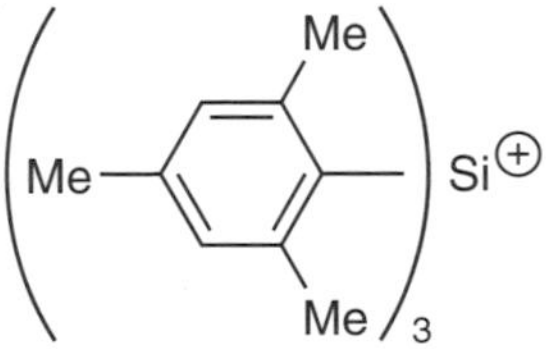

Trimesitylsilylium
Trimesitylsilyl cation

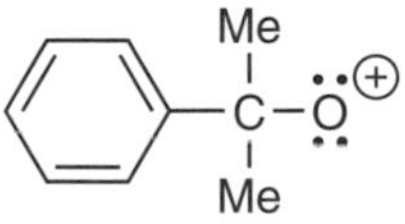

α,α-Dimethylbenzyloxidanylium
(traditional name: ...benzyloxenium)

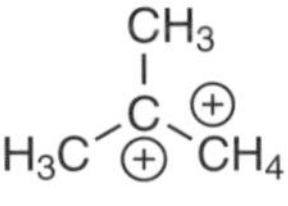

2-Methylpropan-2-ylium-1-ium

Cationic intermediates of defined constitution (π-complexes) such as occur during electrophilic substitution of aromatic compounds are no longer aromatics and should therefore **not** be termed **arenium**- but **...ylium** ions. If there is no fundamental preference for an arbitrarily chosen resonance form (where possible that with the greatest weight) delocalization of charge (and also of π bonds) can be accounted for simply by leaving out all locants.

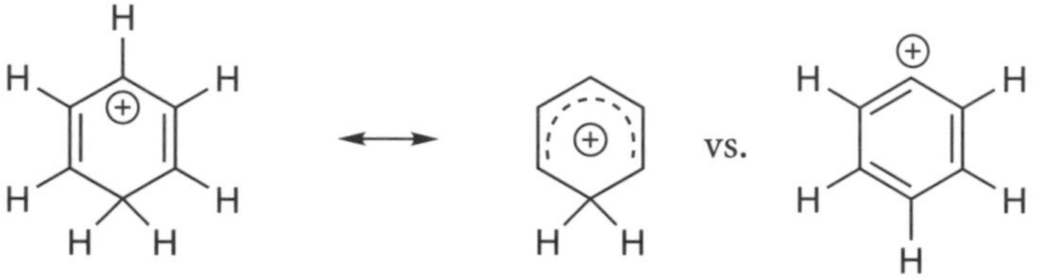

Cyclohexa-2,5-   Cyclohexa-   Phenylium
dien-1-ylium   dienylium   (Benzenylium)

In cases such as the following one, no possibility yet exists of assigning a consistent name for the delocalized system; naming must again be arbitrarily based on the most plausible resonance structure.

1-Methylcyclohexa-   3-Methylcyclohexa-
2,5-dien-1-ylium   2,4-dien-1-ylium

(Locant-1- can be neglected here since the center of charge is always given the lowest locant possible.)

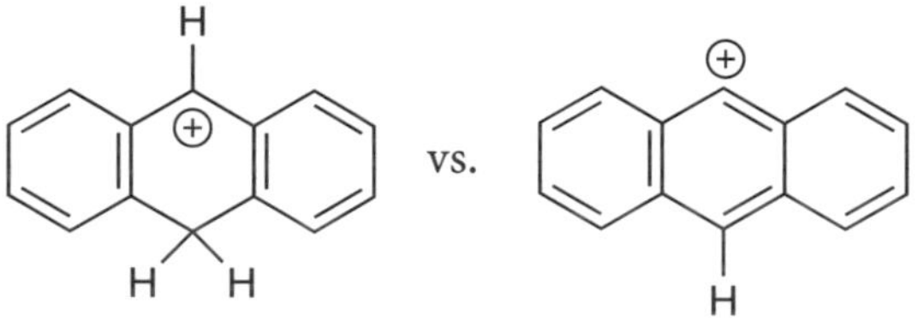

9,10-Dihydroanthra-   9-Anthrylium
cen-9-ylium   (Anthracen-9-ylium)

For cationic aromatics too, naming can be based either on specific resonance forms or, in a less explicit way, on the delocalized system as a whole by omitting any locants.

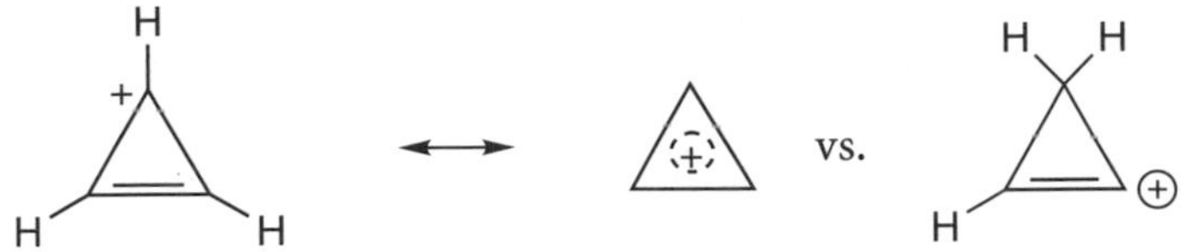

Cycloprop-2-en-
1-ylium

Cycloprop-
enylium

vs.

Cycloprop-1-
en-1-ylium

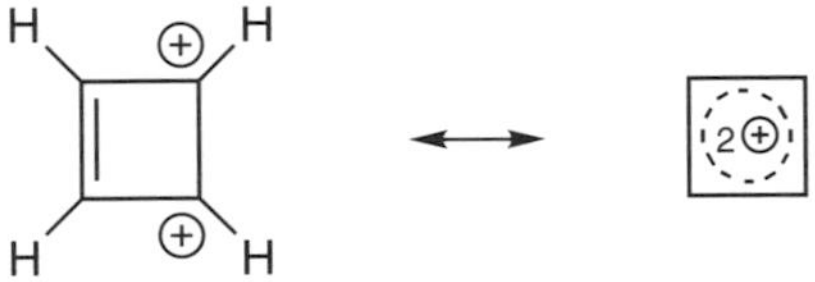

Cyclobut-3-ene-
1,2-di(ylium)

Cyclobutene-
di(ylium)

Particularly for delocalized cations derived from complex heterocyclic aromatics, significant names can sometimes **only** obtained with reference to specifically selected resonance forms.

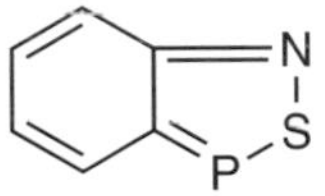

2,1,3-Benzothi-
azaphosphole

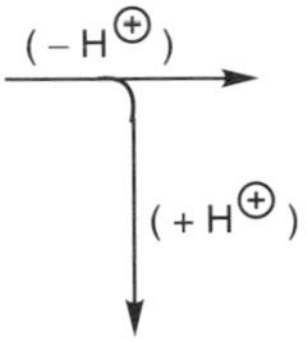

$(-H^{\oplus})$

$(+H^{\oplus})$

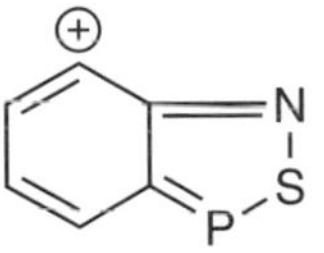

2,1,3-Benzothiaza-phosphol-7-ylium
(localized!)

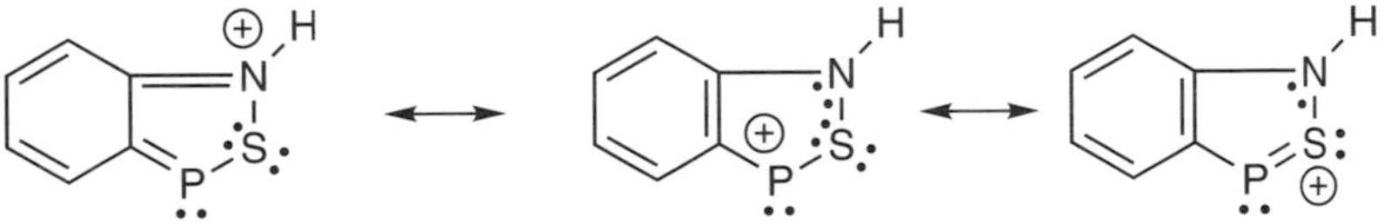

2,1,3-Benzothiaza-
phosphol-1-ium

1,3-Dihydro-2,1,3-
benzothiazaphos-
phol-3-ylium

1,2-Dihydro
-2λ^4,1,3-benzo-
thiazaphos-
phol-2-ylium

Intuitively, one is inclined to opt for the formal 10 π-aromatic resonance form shown on the left.

2.2.7.3
Radical Cations (Cation Radicals)

Formation of these species can be envisaged as resulting from a) removal of an electron from a neutral structure, b) addition of a hydron (or another positively charged group) to a radical, c) addition of a hydron to, accompagnied by removal of a hydrogen atom from, a neutral compound, and d) concomitant abstraction of a hydride ion and a hydrogen atom from a neutral structure (often after a formal dihydrogenation step). Since no discrete morphemes denoting subtraction (or addition) of single electrons have so far been awarded official status, operations c) and d) are generally applied. This translates into appending composite suffixes, such as ...iumyl and ...yliumyl (derivable from Table 11) to the parent name with inclusion, if necessary, of the respective locants. Since here again, in general, and for resonance-prone π systems in particular, only arbitrarily chosen (plausible) resonance forms can be named precisely, the more descriptive denotations **radical cation** (or **cation radical**) – in Chem. Abstr. format: **radical ion(1+)** – are frequently utilized.

$[CH_4]^{•⊕}$ Methaniumyl $[C_{14}H_{10}]^{•⊕}$
         Methane radical cation   Anthraceniumyl etc.
         Methane radical ion(1+)

9,10-Dihydroanthracen-10-ylium-9-yl

$[CH_2]^{•⊕}$ Methyliumyl, λ^2-methaniumyl, methylene radical ion(1+)

trad.: Triphenylammoniumyl
(Triphenylamine radical cation)

syst.: Triphenylazaniumyl
(Triphenylazane radical cation)

Chem. Abstr.: N,N-Diphenylbenzenamine radical ion(1+)

trad.: 4-(Diphenylammoniumylidene)cyclohexa-2,5-dien-1-yl
syst.: 4-(Diphenylazaniumylidene)cyclohexa-2,5-dien-1-yl

That a strictly systematic application of the operational affixes of Table 11 also allows conclusive naming of hitherto neglected structural variants of the **radical cation** types discussed at the beginning of this section is demonstrated by two final examples

$\overset{\oplus}{H_4C}-\overset{\bullet}{CH_2}$

Ethan-2-ium-1-yl

Anthracenium-2-yl

2.2.7.4
Anions

Anions formed by loss of a hydron (proton) from an acid group are named by transforming the end term ... **acid** into ... **oate** or ... **ate**, the term ... **carboxylic acid** into ... **carboxylate**, e.g.: decanoate, naphthalenesulfonate, cyclobutanecarboxylate.

Salts of trivially named formic acid, acetic acid, and butyric acid are called formate, acetate, and butyrate, respectively. Salts of alcohols and thioalcohols are mostly still identified with the traditional name endings ... **olate** (or ... **oxide**) and ... **thiolate** (or ... **sulfide**). Systematically these compounds are to be named as ... **yloxidanides** and ... **ylsulfanides.**

$C_6H_5\overset{..}{\underset{..}{O}}:^{\ominus}$ Phenolate
           Phenoxide
           Phenyloxidanide

$(H_3C)_3C-\overset{..}{\underset{..}{O}}:^{\ominus}$ *tert*-Butoxide
           2-Methylpropan-2-olate
           tert-Butyloxidanide

$H_3C-\overset{..}{\underset{..}{S}}:^{\ominus}$ Methanethiolate, methylsulfide, methylsulfanide

Anions produced by dehydronation (deprotonation) of parent hydrides are characterized by the name ending ... **ide.** More descriptively such anions are denoted as ... **yl anions.**

$H_3C-CH_2-CH_2-\overset{..}{C}H^{\ominus}-CH_3$ Pentan-2-ide, 1-methylbutyl anion!

$:C\equiv C:^{\ominus}$ Ethynediide, ethynediyl dianion (traditional name: acetylide)

Naphthalene-2-ide, 2-Naphthyl anion

$H_2\overset{\ominus}{\overset{..}{C}}-CH_2CH_2CH_2C\equiv C:^{\ominus}$ Hex-1-yne-1,6-diide

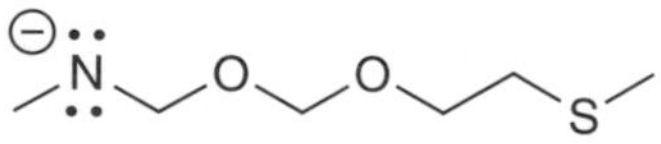

4,6-Dioxa-9-thia-2-azadecan-2-ide

Delocalized anion systems are treated in the same manner as their cationic counterparts: naming either refers in detail to the weightiest and/or most plausible resonance structure or reflects only summarily all functions in question, but without locants.

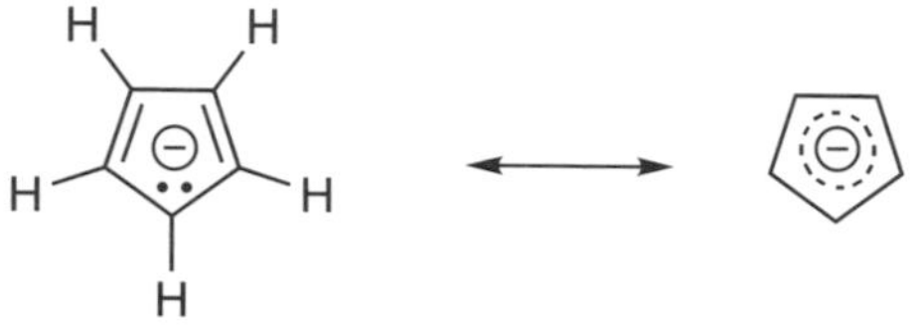

Cyclopenta-2,4-dien-1-ide

Cyclopentadienide
(Cyclopentadienyl anion)

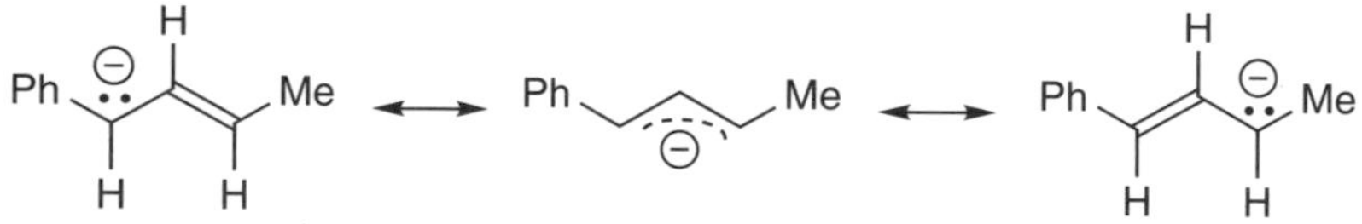

1-Phenylbut-
2-en-1-ide

1-Phenyl-
butenide
(1-Phenylbutenyl anion)

4-Phenylbut-
3-en-2-ide

4H-1,4-Oxazin-4-ide

Finally it should be noted that Chem. Abstr. indexes anions frequently, but by no means always, under the name of the respective parent system and in the rather descriptive format ...**ion(x–)**.

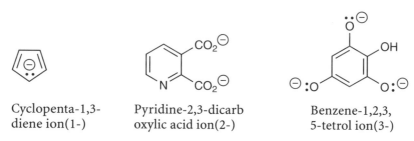

Cyclopenta-1,3-
diene ion(1-)

Pyridine-2,3-dicarb
oxylic acid ion(2-)

Benzene-1,2,3,
5-tetrol ion(3-)

2.2.7.5
Radical Anions (Anion Radicals)

These species can again be derived in four different ways by:

a) Addition of an electron to a neutral structure,
b) subtraction of a hydron from a radical,
c) concomitant addition of a hydride ion to and abstraction of a hydrogen atom from a parent structure,
d) simultaneous abstraction of a hydron and a hydrogen atom from a parent structure, occasionally involving, as in the case of arenes, a formal dihydrogenation.

Naming can then occur either descriptively as ...**radical anion** (a) or strictly operationally with suffix combinations such as ...**uidyl** (c) and ...**idyl** (d). Designations reflecting formation mode b) would lead to the suffix combination ...**ylide** which, regrettably, has already been earmarked as class name for **Wittig reagents** and must therefore be avoided here (see 2.2.7.7). **Delocalization** is treated exactly as in the case of analogous radical cations.

$[C_2H_4]^{\ominus\ominus}$ $\leftrightarrow$ $H_2\overset{\bullet}{C}-\overset{\bullet\bullet}{C}H_2^{\ominus}$ vgl. $H_3C-\overset{\bullet\bullet}{C}H^{\ominus}_{\bullet}$

a) Ethene radical anion
c) Ethenuidyl
d) Ethanidyl   Ethan-2-id-1-yl   Ethan-1-id-1-yl

$[H_3C-CH_3]^{\ominus\ominus}$ vs. $H_4\overset{\ominus}{C}-\overset{\bullet}{C}H_2$ $\overset{\bullet\bullet}{\underset{\bullet}{C}}H_2^{\ominus}$

a) Ethane radical anion   a) λ^2-Methane radical anion
c) Ethanuidyl   Ethan-2-uid-1-yl   c) λ^2-Methanuidyl
                                                        d) Methanidyl

Pyridine radical anion   1,4-Dihydropyridin-1-id-4-yl

With reference to Table 11, more complex radical ions, for example of the **ketyl** type, can be named systematically without difficulties.

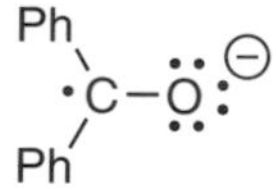

a) Benzophenone   p-Benzoquinone   idem
   radical anion   radical anion

d) Oxidodiphen-   1-Oxido-4-oxo-   4-Oxidophen-
   ylmethyl or: cyclohexa-2,5-   yloxyl or:
   Oxidanidyldi-   dienyl or: 4-Oxidanidyl-
   phenylmethyl   1-Oxidanidyl...   phenyloxidanyl

2.2.7.6
Compounds with Two (or More) Identically Charged Centers

If one and the same compound contains two (or more) cationic or anionic centers, some of these structural elements must be expressed as prefixes in the final name. Cation names ending with ...ium are traditionally prefixed as ...io... and systematically as ...iumyl...; anion names ending with ...ate or ...ide transform into the prefix formats ...ato... and ...idyl..., respectively. Priorities are assigned as follows; cations:

1. C > N > P > As > Sb > Bi > O > S > Se > Te > F > Cl > Br > I
2. Cyclic take precedence over acyclic systems

Anions: order of priority according to Table 7.

$$[(CH_3)_2\overset{\oplus}{Se}-(CH_2)_{12}-\overset{\oplus}{P}(C_6H_5)_3]\ 2\ Cl^{\ominus}$$

(12-Dimethylselenoniododecyl)triphenylphosphonium dichloride
Systematic name: (12-Dimethylselaniumyldodecyl)triphenylphosphanium....

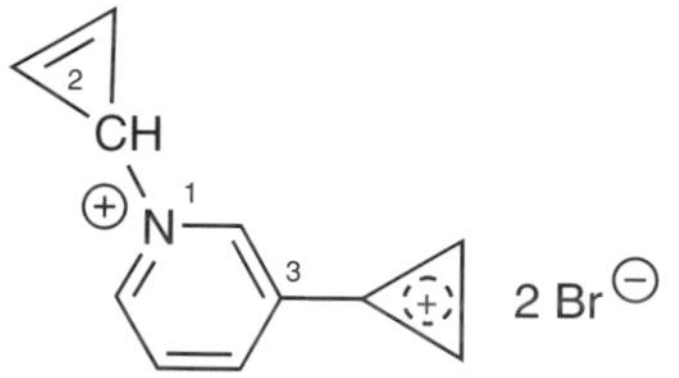

[1-(Cycloprop-2-enyl)pyridinium-4-yl]cyclopropenylium dibromide

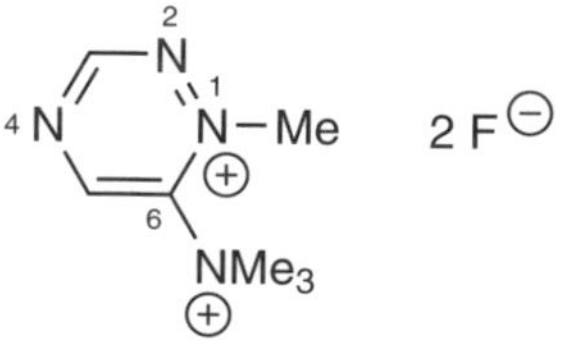

1-Methyl-6-trimethylammonio-1,2,4-triazinium difluoride
system.: ...-6-trimethylazaniumyl-...

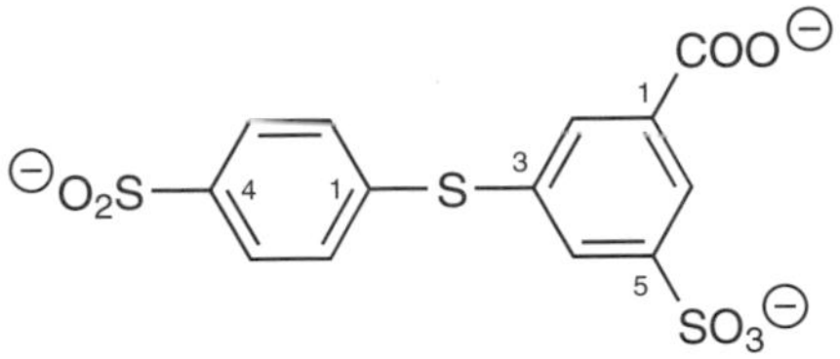

Trisodium 3-(4-sulfinato-
phenylsulfanyl-5-sulfonato-
benzoate

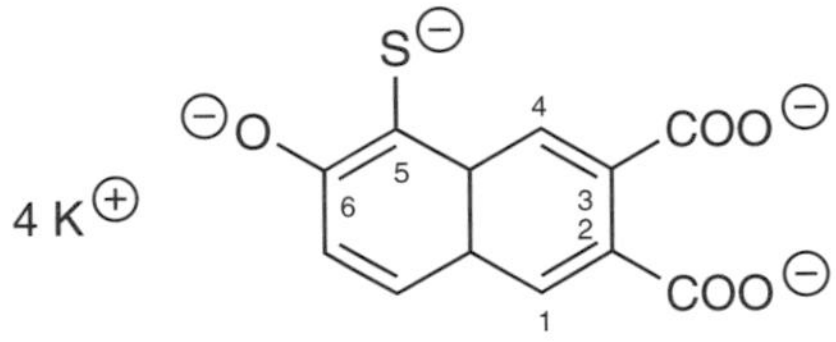

Tetrapotassium 6-oxido-5-
sulfidonaphthalene-2,3-
dicarboxylate
system.: ... 6-Oxidanidyl-5-
sulfanidyl...

2.2.7.7
Compounds with Positively and Negatively Charged Centers (Zwitterions)

If the centers of unlike charge are located in one and the same parent
hydride skeleton the appropriate suffixes of Table 11 are simply used in an
additive manner.

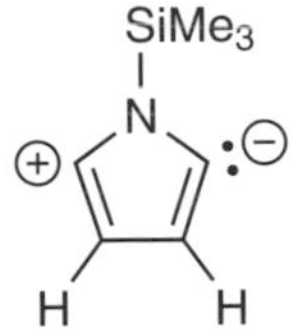

1-Trimethylsilylpyr-
rol-5-ylium-2-ide

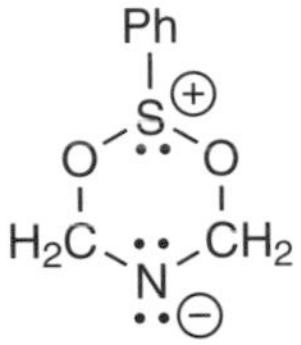

1-Phenyl-1,3,2,5-Dioxa-
thiazinan-2-ium-5-ide

Generally, though, three different naming methods exist for such compounds, depending on how the formal generation of the zwitterion is
envisaged: a) a cation is quasi-substituted in an anion, b) an anion is
attached to a cation (characterized by ... ium or ... ylium), c) a paraphras-
tic symbolization such as **"inner salt"** is applied, as preferred by Chem.
Abstr.

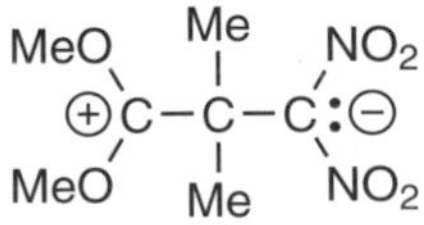

a) 2-(Dimethoxymethyliumyl)-2-methyl-1,1-
dinitropropan-1-ide; 3,3-Dimethoxy-
2,2-dimethyl-1,1-dinitropropan-3-ylium-1-ide

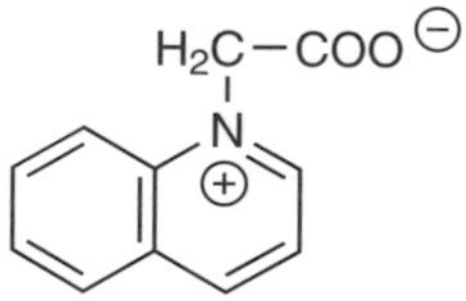

a) (1-Quinolinio)acetate
system.: (Quinolin-1-iumyl)...
Chem. Abstr.: 1-(Carboxy-
methyl)quinolinium hydroxide,
inner salt

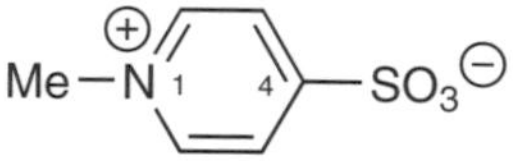

b) 1-Methylpyridinium-4-
sulfonate
CA: 1-Methyl-4-sulfopyrid-
inium hydroxide, inner salt

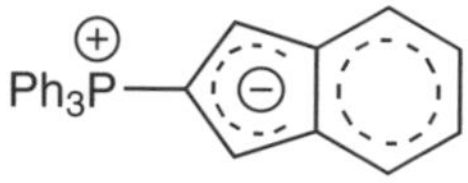

a) 2-(Triphenylphosphonio)indenide
syst.: ...phosphaniumyl)...
trivially: 2-(Triphenylphos-
phonium) indenylide

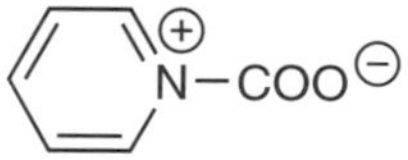

a) (1-Pyridinio)formiate
syst.: (Pyridin-1-iumyl)formiate
b) Pyridinium-1-carboxylate
CA: 1-Carboxypyridinium-
hydroxide, inner salt

Replacement nomenclature can clearly also be utilized for naming zwit-
terions properly, although this approach may appear rather obsolete in
view of the universally applicable directions of Table 11.

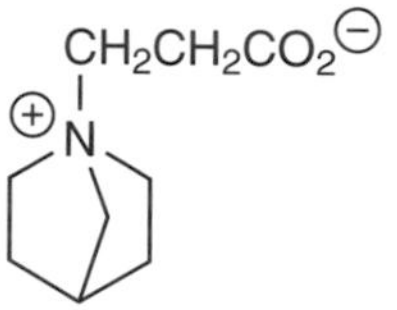

3-(1-Azoniabicyclo[2.2.1]heptan-
1-yl)propanoate
a) 3-(1-Azabicyclo[2.2.1]heptan-
1-iumyl)...

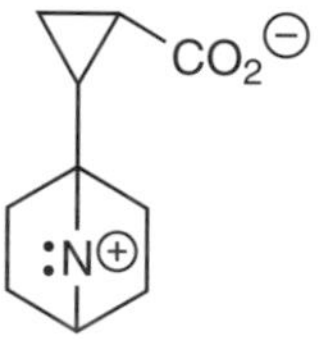

2-(7-Azanyliabicyclo[2.2.1]heptan-
1-yl)cyclopropanecarb-
oxylate
a) 2-(7-Aza...-7-ylium-1-yl)...

Chem. Abstr. will certainly again use "inner salt" names for the last two
examples.

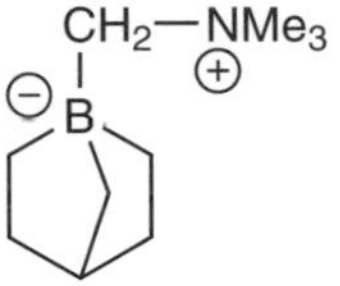

1-(Trimethylammoniomethyl)-
1-boratabicyclo[2.2.1]heptane

a) 1-(Trimethylazaniumylmethyl)-
1-borabicyclo[2.2.1]heptan-1-uide

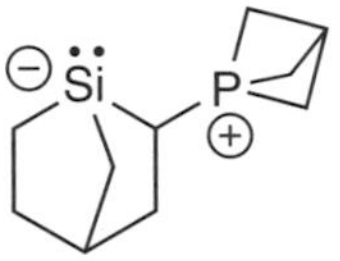

2-(1-Phosphoniabicyclo[1.1.1]pentan-1-yl)-
1-silanidabicyclo[2.2.1]heptane

a) 2-(1λ^5-Phosphabicyclo[1.1.1]pentan-1-
yliumyl)-1-siliabicyclo[2.2.1]heptan-1-ide

As before, delocalized systems can be named reasonably only on the basis of their weightiest and/or most plausible resonance structure.

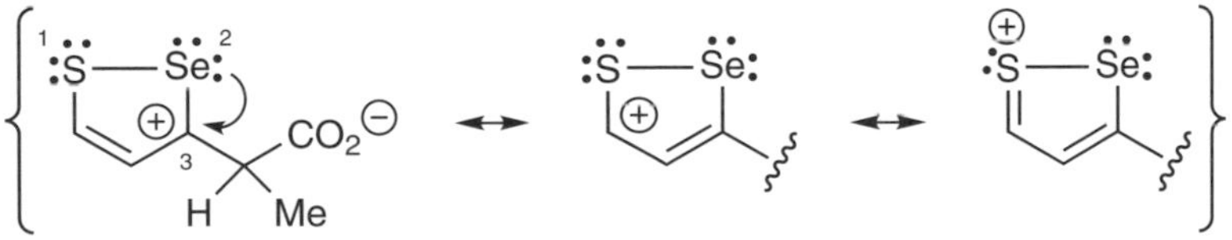

a) 2-(3H-1,2-Thiaselenol-3-ylium)propanoate
CA: 3-(1-Carboxyethyl)-1,2-thiaselenolium
hydroxide, inner salt

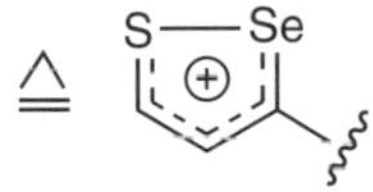

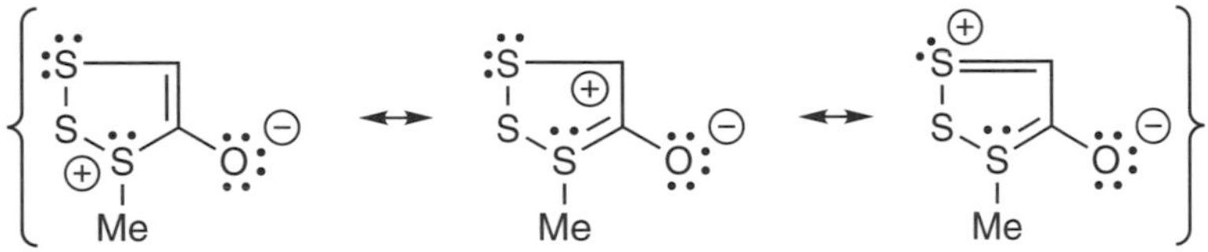

a) 3-Methyl-1,2,3-trithiolium-4-ylox(**idan**)ide
b) 3-Methyl-1,2,3-trithiolium-4-olate
CA: 5-Hydroxy-1-methyl-1,2,3-trithiolium
hydroxide, inner salt

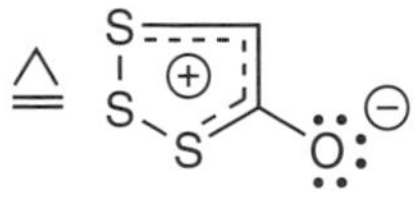

3 Brief Demonstration
of the General Nomenclature Rules
for the Most Important Traditional Compound
Classes (Functional Parents)

Because the use of different kinds of nomenclature for one and the same compound class is still widespread, it is necessary to bring into focus which principles are to be applied in general. However, only those compound classes will be discussed individually whose naming does not axiomatically follow from Table 8. As a rule, frequently occurring **trivial names** are compiled in tables in the appendix. The remaining compound classes are mentioned in the examples. Incidentally, it should be noted once again that in the context of **substitutive nomenclature** certain class names no longer appear at all in the names of the pertinent individual compounds, as in the case of alcohols, ethers, ketones, etc.

3.1
Carboxylic Acids, Sulfonic Acids, etc. and their Derivatives

Substitutive nomenclature allows two naming alternatives for carboxylic acids, depending on whether the carbon atom of the functional group is included in the parent name or not included. For linear acids the first alternative is generally preferred if no more than two carboxylic groups are present. For cyclic carboxylic acids in all cases the whole functional group is expressed as a suffix.

$$\overset{6}{H_2C}=\overset{5}{C}=\overset{4}{CH}-\overset{3}{\underset{\underset{Ph}{|}}{\overset{\overset{OH}{|}}{CH}}}-\overset{2}{\underset{|}{CH}}-\overset{1}{COOH}$$

3-Hydroxy-2-phenyl hexa-4,5-dienoic acid

$$\overset{1}{\underset{\overset{||}{O}}{H_2N-\overset{||}{C}}}-\overset{2}{\underset{\underset{O}{||}}{C}}-\overset{3}{C}\equiv C-\overset{5}{CH}=CH_2-\overset{7}{\underset{|}{\overset{\overset{HCO}{|}}{CH}}}-\overset{8}{\underset{\underset{NH_2}{|}}{CH}}-\overset{9}{\underset{\overset{||}{O}}{C}}-NH_2$$

8-Amino-7-formyl-2-oxonon-5-en-3-ynediamide

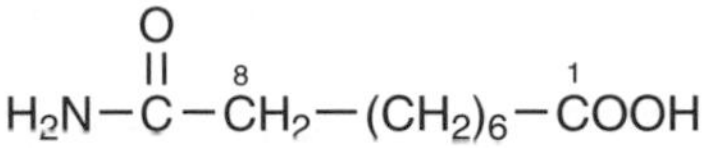

$$\text{H}_2\text{N}-\overset{\overset{\displaystyle O}{\|}}{\text{C}}-\overset{8}{\text{CH}_2}-(\text{CH}_2)_6-\overset{1}{\text{COOH}}$$ 8-Aminocarbonyloctanoic acid

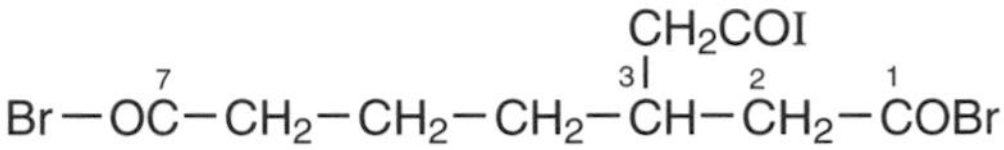

$$\text{Br}-\overset{7}{\text{OC}}-\text{CH}_2-\text{CH}_2-\text{CH}_2-\overset{\overset{\displaystyle \overset{3}{\text{CH}_2}\overset{2}{\text{CO}}\overset{1}{\text{I}}}{|}}{\text{CH}}-\text{CH}_2-\text{COBr}$$ 3-(Iodocarbonylmethyl)-heptanedioyl dibromide

In the meantime Chem. Abstr. has adopted a completely systematic treatment also for side chains, which, in the above example, leads to 3-(2-iodo-2-oxoethyl)… for the substituent group.

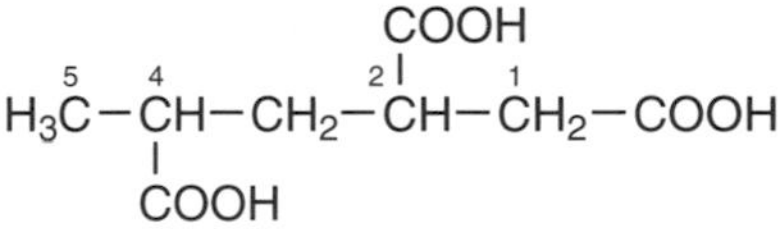

$$\text{H}_3\text{C}-\overset{4}{\underset{\underset{\displaystyle \text{COOH}}{|}}{\text{CH}}}-\text{CH}_2-\overset{2}{\underset{\overset{\displaystyle \text{COOH}}{|}}{\text{CH}}}-\overset{1}{\text{CH}_2}-\text{COOH}$$ Pentane-1,2,4-tricarboxylic acid

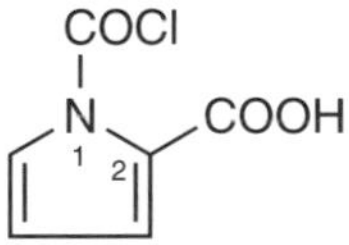

1-(Chlorcarbonyl)- resp. 1-(Chloroxomethyl)pyrrole-2-carboxylic acid

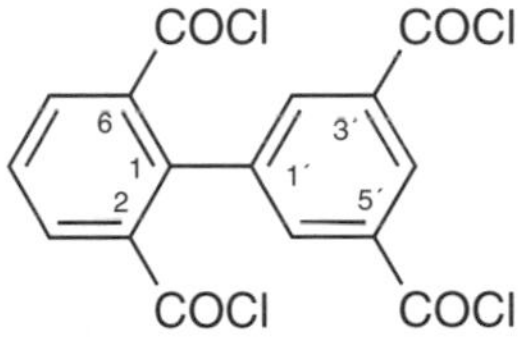

Biphenyl-2,6,3',5'-tetracarbonyl tetrachloride

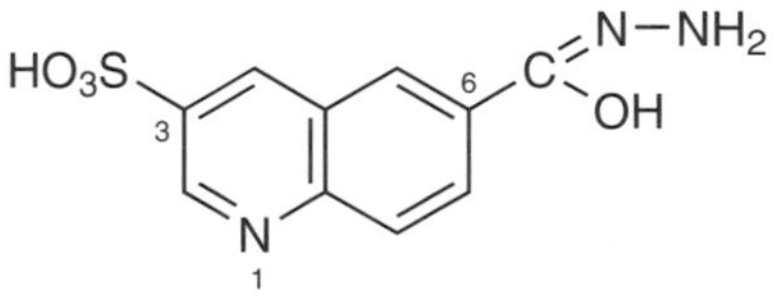

3-Sulfoquinoline-6-carbo-hydrazoic acid

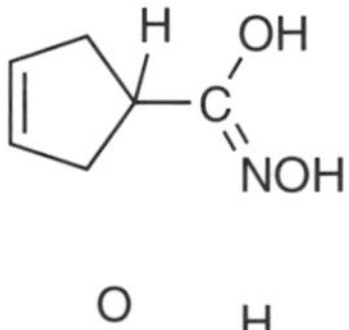

Cyclopent-3-enecarbohydroximic acid
CA: N-Hydroxycyclopent-3-enecarboximidic acid

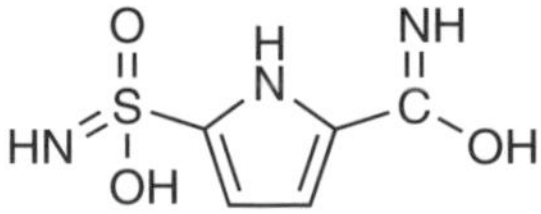

5-[Hydroxy(imino)sulfinyl]-1H-pyrrole-2-carboximidic acid

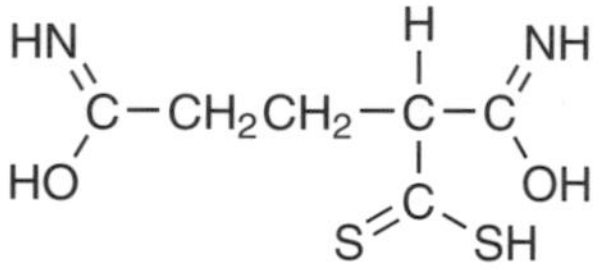

2-(Dithiocarboxy)pentanedi(imidic acid)

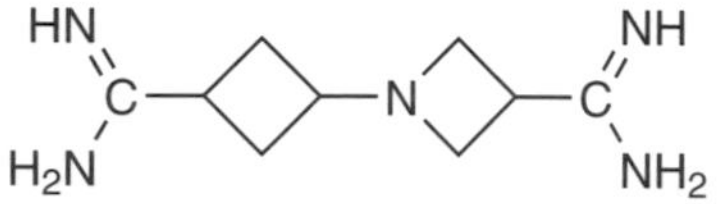

1-(3-Amidinocyclobutyl)azetidine-3-carboxamidine
CA: 1-[3-(Aminoiminomethyl)cyclobutyl]azetidine-3-carboximidamide

For trivially named thioacids and their Se and Te analogues the following convention holds: (di, tri)thiocarbonic, thioformic, thioacetic, thiobenzoic, etc., acid; systematic treatment of such acids follows from Table 8.

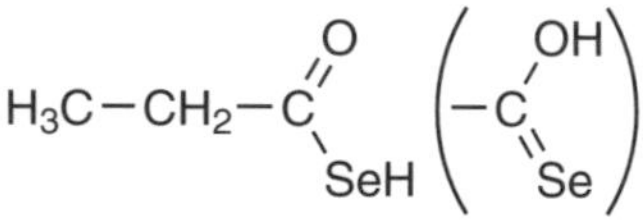

Propaneselenoic acid
(Propaneselenoic O-acid)

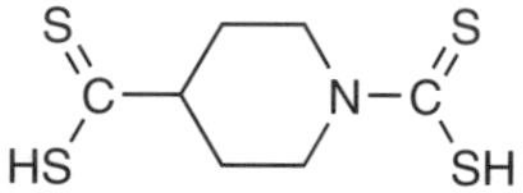

Piperidine-1,4-bis(carbodithioic acid)

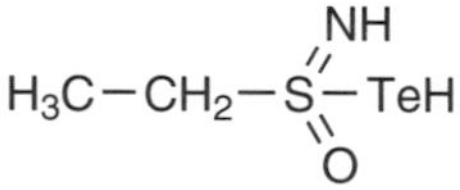

Ethanesulfonimidotelluroic acid

In a similar manner "peracids" can be named trivially or systematically: per(or peroxy)carbonic, performic etc. acid.

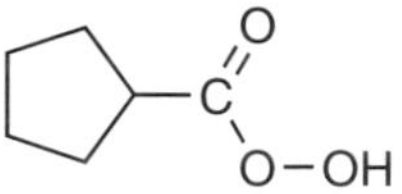

Cyclopentaneperoxycarboxylic acid
CA: Cyclopentanecarboperoxoic acid

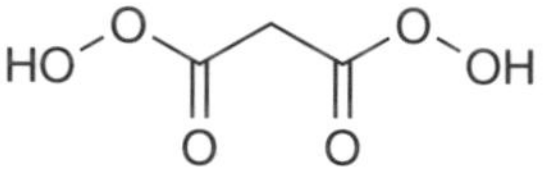

Propanedi(peroxoic acid)
or Diperoxypropanedioic acid

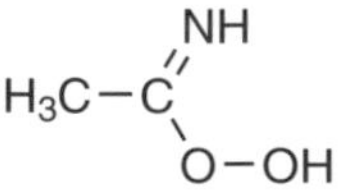

Ethanimidoperoxoic acid
triv.: Peroxyacetimidic acid

Conjunctive nomenclature is particularly suitable for systems multiply substituted with identical functional parents but is generally used by Chem. Abstr. for any type of saturated linear monocarboxylic acid bearing a ring as substituent group.

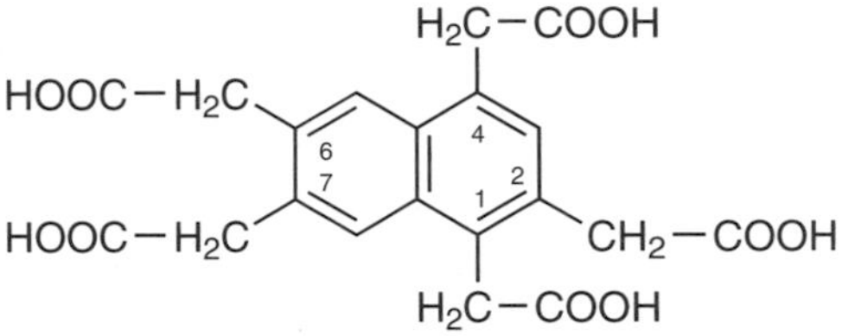

conjunct.: Naphthalene-1,2,4,6,7-pentaacetic acid,
substit.: Naphthalene-1,2,4,6,7-pentaylpentaacetic acid

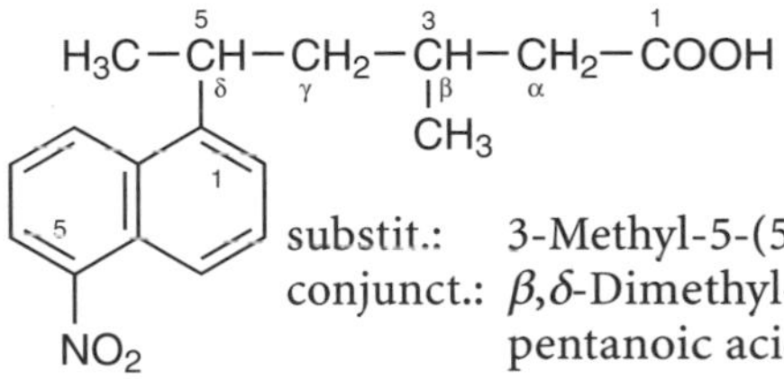

substit.: 3-Methyl-5-(5-nitro-1-naphthyl)hexanoic acid
conjunct.: β,δ-Dimethyl-5-nitronaphthalene-1-pentanoic acid

In an exceptional ruling hemiamides and hemialdehydes of trivially named carboxylic acids can be designated as **…amidic and aldehydic acids,** respectively.

OHC–CH$_2$–COOH
Malonaldehydic acid,
syst.: 3-Oxopropanoic acid

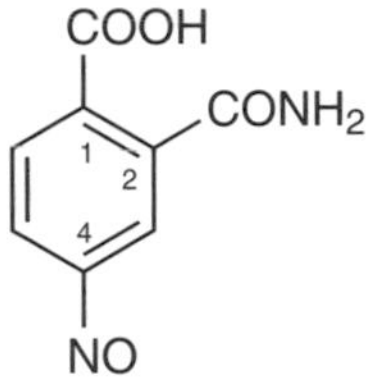

4-Nitrosophthalamic acid
syst.: 2-Aminocarbonyl-4-nitrosobenzoic acid

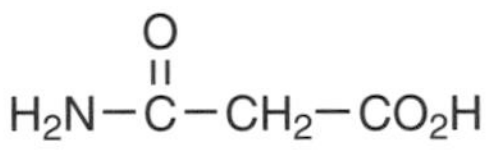

$$H_2N-\overset{\overset{\displaystyle O}{\|}}{C}-CH_2-CO_2H$$
Malonamidic acid, syst.: 3-Amino-3-oxopropanoic acid

Substituent groups ("radicals") derived from acids and certain derivatives thereof by removal of –OH from the functional group are generally called **acyl groups** and individually named by transforming the ending **…ic acid** to **…yl** or **…oyl** and **…carboxylic acid** to **…carbonyl.** The names thus formed can also be used in "radicofunctional" (functional class) nomenclature.

$$H_2C \underset{\overset{C}{H_2}}{\overset{\overset{C}{H_2}}{\diamondsuit}} CH-CO-\xi \qquad \text{Cyclobutanecarbonyl}$$

The above group name is used only in its function as acyl group, e.g. for acyl halides etc.; in all other cases it must read: cyclobutylcarbonyl!

$$\xi-\overset{O}{\overset{\|}{C}}-C\equiv C-CH_2-CH=C=CH-\overset{O}{\overset{\|}{C}}-\xi$$

Octa-2,3-diene-6-ynedioyl
Systematic name: 1,8-Dioxo-octa-2,3-diene-6-yne-1,8-dioyl

$$\xi-OC \qquad CO-\xi$$
$$\xi-OC-\overset{3\ \ 2}{\underset{4\ \ \ 1N}{\bigcirc}}$$

Pyridine-2,3,4-tricarbonyl

$$\xi-\overset{7}{OC}-\overset{6}{CH}-CH_2-\overset{4}{CH}-CH_2-\overset{2}{CH}-\overset{1}{CO}-\xi$$
$$\quad\ \ \ |\qquad\qquad |\qquad\qquad |$$
$$\quad\ \ CH_3\qquad\ \ COOH\quad\ \ COOH$$

2,4-Dicarboxy-6-methylheptanedioyl
syst.: 1,7-Dioxo-2,4-di...heptane-1,7-diyl

$$H_3C-(CH_2)_{10}-\overset{\overset{NH}{\diagup}}{C}$$

Dodecanimidoyl
syst.: 1-Iminododecyl

Salts and esters of carboxylic acids and related compounds are named by placing the name of the metal or the esterifying group in front of the name of the acid anion, separated by a space. Joint occurrence of both functions is indicated accordingly. Acid salts and esters are named analogously. Complex cases can also be named more descriptively: ...salt or ...ester of ...acid. (Similar provisions are valid for other acid derivatives such as amides and nitriles.)

$$H_3C-(CH_2)_6-COOLi$$ Lithium octanoate (or lithium salt of octanoic acid)

$$\overset{COOMe}{\underset{|}{N\diagdown N}}$$

Methylpyrazole-1-carboxylate (or Pyrazole-1-carboxylic acid methyl ester)

$$HON\underset{\ominus O}{\overset{\diagdown}{C}}-(CH_2)_7-\underset{O^\ominus}{\overset{NOH}{\overset{\diagup}{C}}}\qquad Na^\oplus,\quad H^\oplus$$

Sodium hydrogen nonanedihydroximate

$$\begin{array}{l} H_2C-COO^\ominus \\ |\\ HC-COOEt \qquad K^\oplus,\quad H^\oplus \\ |\\ H_2C-COO^\ominus \end{array}$$

Potassium 2-ethyl hydrogen propane-1,2,3-tricarboxylate (Monopotassium salt of Propane-1,2,3-tricarboxylic acid 2-ethyl ester)

$$Me-COO-\!\!\left\langle\!\!\bigcirc\!\!\right\rangle\!\!-OOC-CH_2Cl$$

1,4-Phenylene acetate chloro-acetate

$$H_3C-CH_2-CH_2-CH_2-CH_2-\underset{OMe}{\overset{OMe}{\overset{|}{C}}}-OMe$$

Trimethyl orthohexanoate
syst.: 1,1,1-Trimethoxyhexane

$$H_3C-C(-OiPr)_3$$

Tetraisopropyl orthoacetate
Systematic name: 2,2′,2″-(Ethylidynetrisoxy)tris-propane

$$H_3C-\underset{OEt}{\overset{NH}{\overset{\diagup}{\underset{\diagdown}{C}}}}$$

syst.: Ethanimidic acid ethyl ester
triv.: Acetimidic acid ethyl ester

$$H_3C-\overset{O}{\overset{\|}{C}}-\underset{Cl}{\overset{NBu}{\overset{\diagup}{\underset{\diagdown}{C}}}}$$

syst.: N-Butyl-2-oxopropanimidoyl chloride
triv.: 2-Oxopropionic acid butylimide chloride

$$\left\langle\!\!\bigcirc\!\!\right\rangle_{N}-\underset{NHMe}{\overset{NMe}{\overset{\diagup}{\underset{\diagdown}{C}}}}$$

syst.: N,N′-Dimethylpyridine-2-carboximidamide
trad.: N,N′-Dimethylpyridine-2-carboxamidine
triv.: Pyridine-2-carboxylic acid methylamide methylimide

If a senior functional group is present ester functions must be represented in the form of substituent group prefixes.

[4-(Cyclobutylcarbonyloxy)-3-(cyclopentyloxycarbonyl)phenyl]diphenylsulfonium iodide

Thiocarboxylic acids, sulf(on, in, en)ic acids, and their derivatives are treated analogously; if necessary specific positions of esterification must be marked separately by placing an indicative letter locant (in italics) in front of the name of the esterifying group.

$H_3C-CH_2-SO_2Br$ Ethanesulfonyl bromide (otherwise ethylsulfonyl!)

4-[(Ethylsulfanyl)carbonyl]-2-[(pentyloxy)thiocarbonyl]furan-3-carbodithioic acid

$^{\ominus}SSC-(CH_2)_3-CSS^{\ominus}Na^{\oplus}, H^{\oplus}$ Sodium hydrogen pentanebis(dithioate)

$H_3C-(CH_2)_7-CO-SC_2H_5$ S-Ethyl nonanethioate

O-Ethyl ethanethiosulfinate

5-[Benzyloxy(thiosulfonyl)]-4-(methylaminosulfinyl)-2-[(methylsulfanyl)sulfinyl]benzenedithiosulfonic S-acid

Although the above principles allow all kinds of esters to be named in a uniform way, for indexing purposes a number of additional points have to be observed. It is, for example, not very sensible to index an ester of acetic acid with a very complex alcohol under the keyword carboxylic acid which is ranked higher according to the seniority order of compound classes. Because of this, Chem. Abstr. evaluates all alcohols and acids according to more or less arbitrarily designed criteria of complexity (for details see Chem. Abstr. Index Guides), on the basis of which it is then to be decided which index entry a certain ester has to be associated with. This **registry name** again has more the character of a parent system-modifying description than that of a name proper. A few examples may serve to illustrate this point.

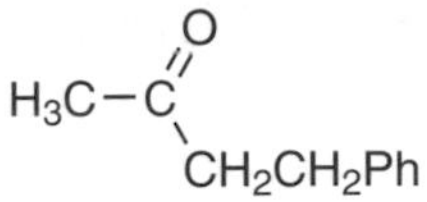

H_3C-C O / CH_2CH_2Ph

Acetic acid, esters,
2-phenylethyl ester

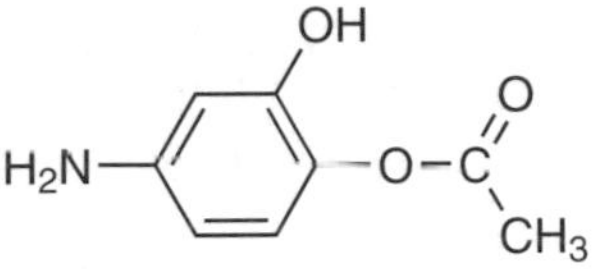

H_2N- OH / $-O-C$ O CH_3

4-Aminobenzene-1,4-
diol, 1-acetate

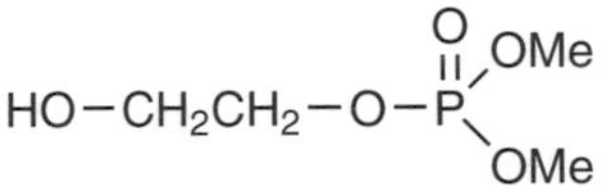

$HO-CH_2CH_2-O-P$ O OMe / OMe

Phosphoric acid, esters,
2-hydroxyethyl dimethyl ester

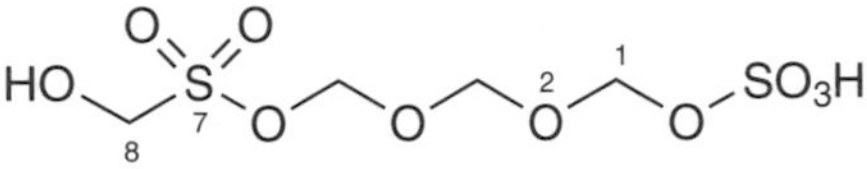

HO O S O $_7$ $_8$ O O $_2$ $_1$ O SO_3H

2,4,6-Trioxa-7-thiaoctane-1,8-diol,
mono(hydrogensulfate) 7,7-dioxide

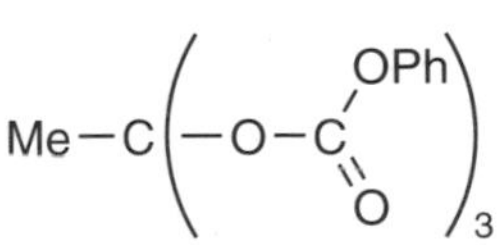

$Me-C$ ($-O-C$ OPh / O) $_3$

Ethane-1,1,1-triol, esters,
tribenzoate

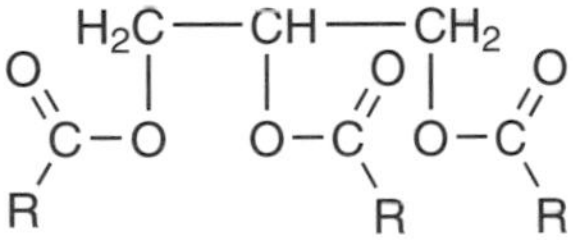

$H_2C-CH-CH_2$
O | | O | O
$C-O$ $O-C$ $O-C$
R R R

$R = C_{17}H_{33}$

Octadecanoic acid, esters,
propane-1,2,3-triyl ester

3.2
Nitriles, Isocyanides, and Similar Compounds

Like carboxylic acids, nitriles can be named substitutively by two methods,
with or without inclusion of the C atom of the functional group in the parent name.

$C_6H_{13}-CN$ Heptanenitrile, $NC-$ H S ^{6} ^{1} N^2 / N ^{3} CN / $_4$

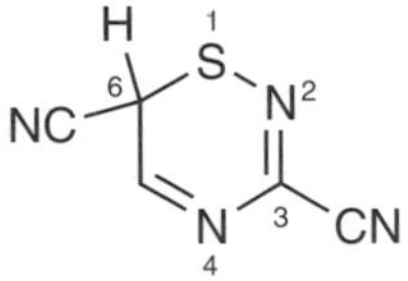

6H-1,2,4-Thiadiazine-
3,6-dicarbonitrile

Nitriles derived from trivially named carboxylic acids traditionally have
the ending ...onitrile: acetonitrile, propiononitrile, butyronitrile, etc.
Moreover, **radicofunctional nomenclature** is (still) also customary here,
leading to names of the type ...**yl cyanides.**

$H_3C-CH_2CH_2-CN$
Propyl cyanide,

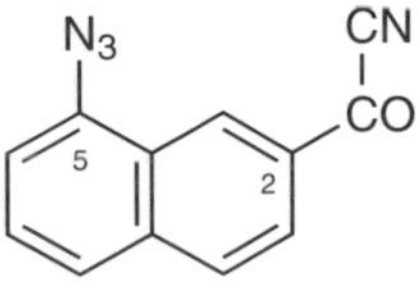

N_3 CN | CO

5-Azido-2-naphthoyl
cyanide
CA: α-Oxo-5-azidonaph-
thalene-2-acetonitrile

Isocyanides and analogous compounds are exclusively named in a radico-functional way, i. e. with functional class designations, provided the functional group is not expressed as a substitutive prefix.

$NCSe-CH_2-CH_2-OCN$ 2-Isoselenocyanatoethyl cyanate
Chem. Abstr.: Cyanic acid 2-isoseleno-cyanatoethyl ester

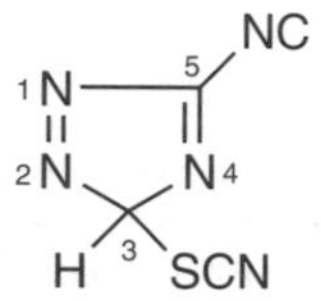

3-Thiocyanato-3H-1,2,4-triazol-5-yl isocyanide
CA: 3-Thiocyanato-5-isocyano-3H-1,2,4-triazole

3.3
Aldehydes and Ketones

The two substitutive naming approaches for aldehydes are used exactly as in the case of carboxylic acides, nitriles, etc.

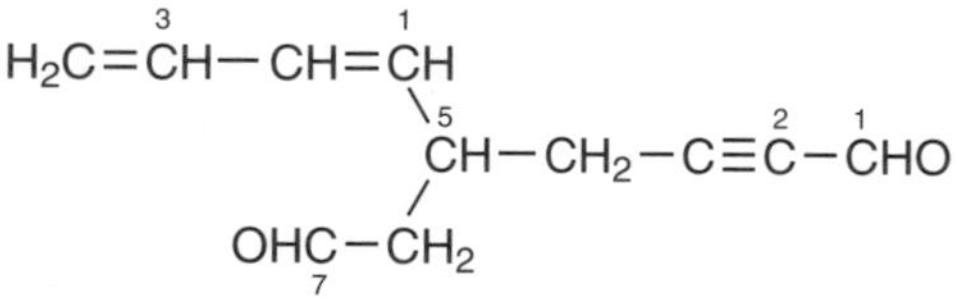

5-(Buta-1,3-dienyl)hept-2-ynedial

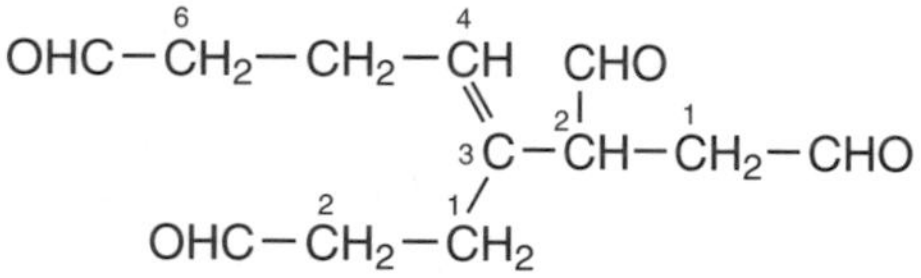

3-(3-Oxopropyl)hex-3-ene-1,2,6-tricarbaldehyde

For linear aldehydes bearing cyclic substituent groups, Chem. Abstr. again employs exclusively conjunctive names.

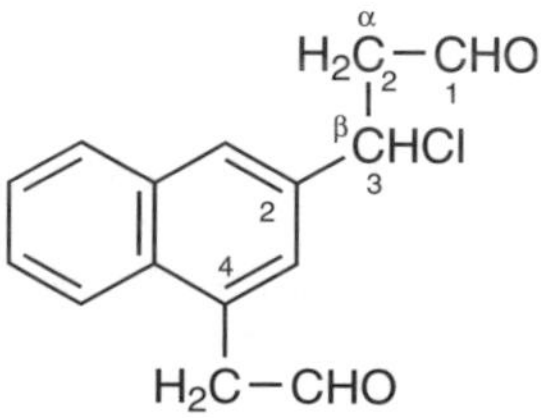

subst.: 3-Chloro-3-[4-(2-Oxoethyl)-2-naphthyl]propanal

conj.: β-Chloro-4-(2-oxoethyl)naphthalene-2-propanal

OHC—H₂C   CH₂—CHO   conj.:   Thiophene-2,3,4-triacetal-
         ╲4   3╱   dehyde
          ║║   ║║   subst.: Thiophene-2,3,4-triyltris-
          ╲1   2╱   acetaldehyde
            S   CH₂—CHO

The possibilities for naming **ketones** are manifold but not stringently delimited with regard to their respective areas of application. Nonetheless, is it again the rules of substitutive nomenclature that are most uniformly applicable here and should therefore be prefered.

$$H_3\overset{11}{C}-CH_2-CH_2-CH_2-\underset{\underset{O}{\|}}{\overset{7}{C}}-\underset{\underset{\underset{\underset{\|}{O}}{C-CH_3}}{CH_2-CH_2-\underset{3}{C}}}{\overset{6}{C}H}-CH_2-CH_2-\overset{3}{\overset{\|}{\overset{O}{C}}}-CH_2-\overset{1}{C}H_3$$

6-(3-Oxobutyl)undecane-3,7-dione

H₂C=C=O   O=C=CH–CH₂–CH=O
Ethenone (ketene)   Penta-1,4-diene-1,5-dione

The substitutive method is also applied in a straightforward manner when the keto function is unilaterally attached to a cyclic component. To prevent breaking any rules, a nomenclatorial/semantic artifice is unavoidable here. This consists in the renaming of an alkan**al**, in principle **not** substitutable at its 1-position, as a fictive, substitutable, alkan-1-**one**. This leads to the

                              subst.:   1-(3H-Indazol-3-yl)hexan-1-one
                              radicofunct.: (3H-Indazol-3-yl)pentylketone
                              (see below)

extreme but stringent inference that ketones with cyclic groups on both sides of the functionality are to be named generally as derivatives of the unrestrictedly substitutable parent system **methanone** (instead of the non-substitutable **methanal**) which are therefore indexed (by Chem. Abstr.) as such. Accordingly, traditional functional class (radicofunctional) names such as diphenyl **ketone** (benzophenone) can be elegantly transformed into substitutive names of the form diphenyl**methanone!**

In the presence of senior functional groups the ketone component must be expressed in the form of an appropriate prefix. Here only the classical formyl, acetyl and benzoyl groups are still indexed as such; all other keto substituents are regarded as *n***-oxoalkyl groups** and named as such.

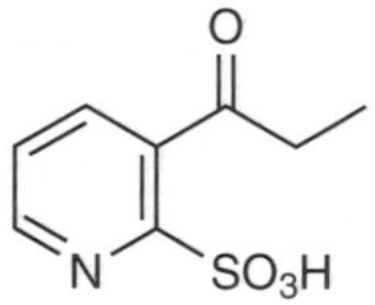

3-(1-Oxoproypl)pyridine-2-sulfonic acid

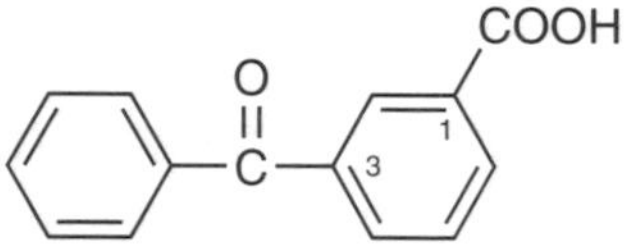

3-Benzoylbenzoic acid

The traditional endings **...ophenone** and **...onaphthone** are still frequently used for ketones of the above type containing phenyl or naphthyl components.

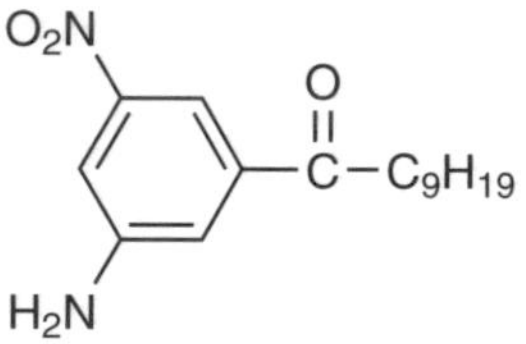

3'-Amino-5'-nitrodecanophenone
substit.: 1-(3-Amino-5-nitrophenyl)decan-1-one
radicof.: (3-Amino-5-nitrophenyl)nonyl ketone (see below)

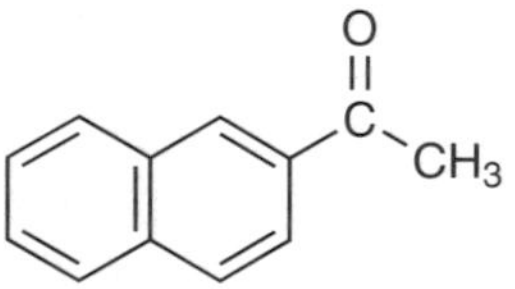

2-Acetonaphthone
substit.: 2-Naphthylethanone
radicof.: Methyl(2-naphthyl)ketone (see below)

The most straightforward method for naming monoketones is certainly provided by **radicofunctional nomenclature** which individualizes the class designation **ketone** by placing it behind the separately cited terms for the respective substituent groups. As has been shown above, though, the **functional class names** obtained in this way can readily be transformed into **substitutive registry names** by redefining **...ketone** as **...methanone.**

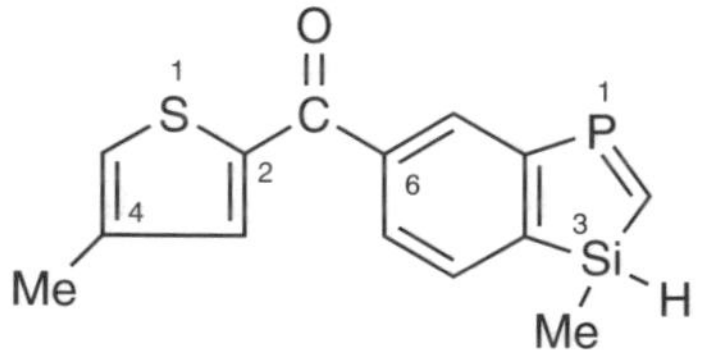

(3-Methyl-3*H*-1,3-benzophospha-silol-6-yl)(4-methyl-2-thienyl)ketone
substit.: ...methanone

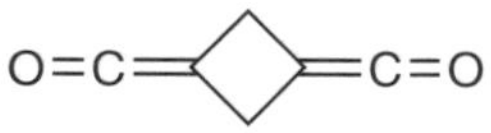

$O=C{=}\langle\ \rangle{=}C=O$ Cyclobutane-1,3-di(ylidene)bismethanone!

Polyketones in which a chain of directly linked keto groups is connected at both ends to rings are named **substitutively** or **radicofunctionally**.

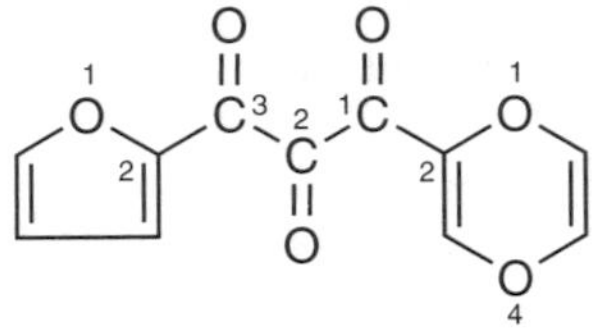

1-(1,4-Dioxin-2-yl)-3-(2-furyl)propane-
trione or:
(1,4-Dioxin-2-yl)(2-furyl)triketone

Cyclic ketones are generally named substitutively. If CH₂ groups are already present in the underlying cyclic parent system or can be generated by formal di-, tetra-, etc. hydrogenation, this procedure is unproblematical. If, however, as a corollary of the transformation of a CH group into a CO group an additional position has to be saturated, this must be accounted for by indication of an **"added" hydrogen.** In such cases the CO groups is, exceptionally, of higher priority than the added hydrogen.

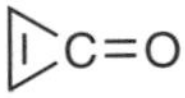

Cyclopropenone

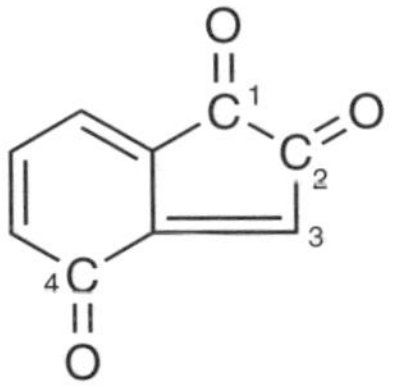

Indene-1,2,4-trione more precise
but with unnecessary redundancies:
1H-Indene-1,2(4H),4-trione

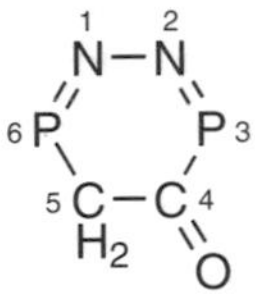

1,2,3,6-Diazadiphosphinin-4(5H)-one
formerly: 1,2,3,6-Diazadiphosphorin-4
(5H)-one

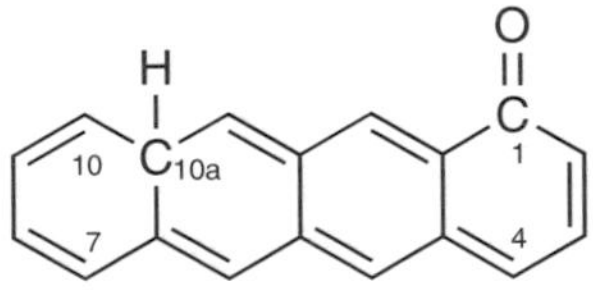

Naphthacen-1(10aH)-one

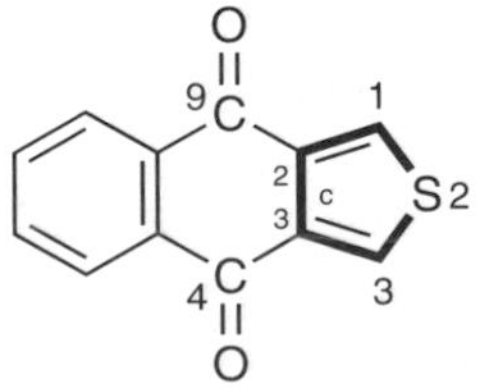

Naphtho[2,3-c]thiophene-4,9-dione

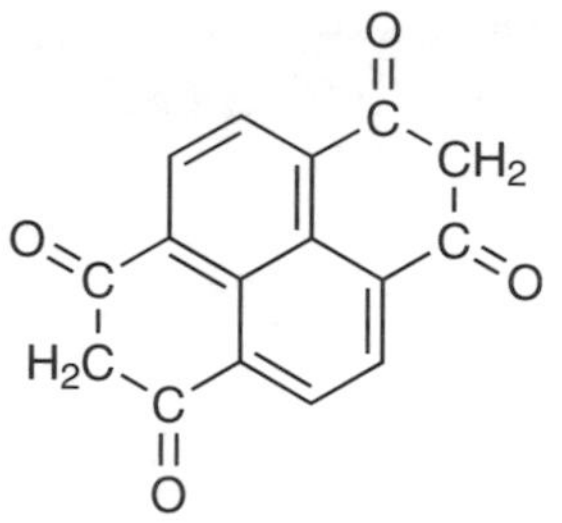

Pyrene-1,3,6,8(2*H*,7*H*)-tetrone

Note:
For a number of ketones derived from heterocycles whose names end with
...**idine** and ...**oline** the abbreviated suffixes ...**idone** and ...**olone** are still
often used instead of the systematic endings ...**idinone** and ...**olinone**,
e.g.: pyridone vs. pyrrolidinone, quinolone vs. quinolinone, piperidone vs.
piperidinone, pyrrolidone vs. pyrrolidinone, etc. It is recommended that
proper systematic names are use here too.

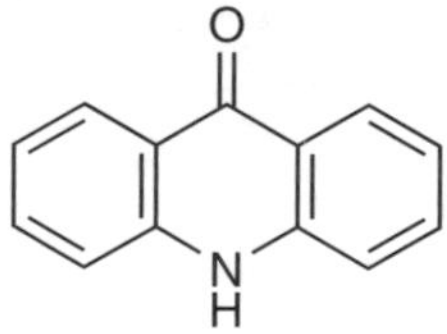

Acridin-9(10*H*)-one
and not: 9-Acridone

Arene-derived di- and tetraketones presenting the typical quinoid double
bond arrangement can be characterized by name suffixes such as
...**quinone**, ...**diquinone** etc. the parent name sometimes being subject to
certain changes.

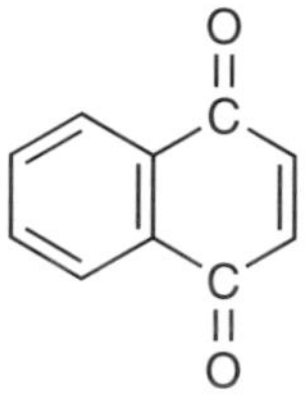

1,4-Naphthoquinone
(Naphthalene-1,4-dione)

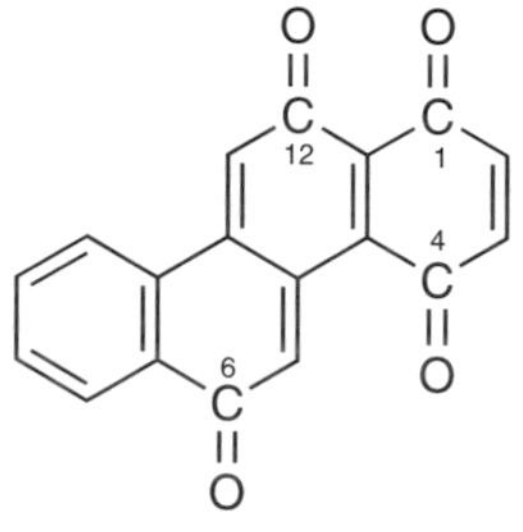

1,4,6,12-Chrysenediquinone
(Chrysene-1,4,6,12-tetrone)

Transcription of these rules for naming **thioaldehydes** and **thioketones**
causes no problems as will be demonstrated by just two examples.

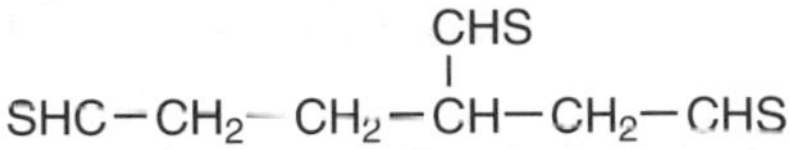

$$CHS$$

SHC$-$CH$_2-$ CH$_2-$CH$-$CH$_2-$CHS   Butane-1,2,4-tricarbothioaldehyde

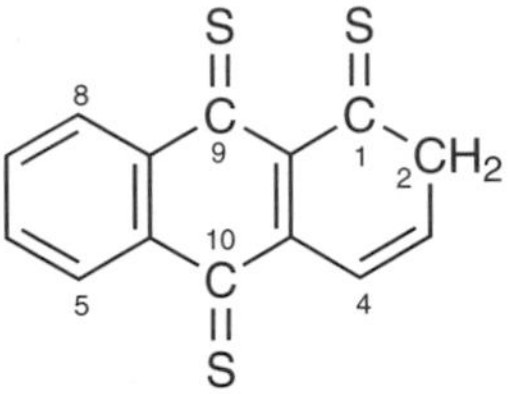

Anthracene-1,9,10(2H)-trithione

Thioaldehydes and thioketones oxidized at sulfur are additively named as **thioaldehyde oxides, dioxides** and **thioketone oxides, dioxides,** respectively. Besides, the traditional names **sulfine** and **sulfene** are still used for these species.

C$_6$H$_5$-CH=SO   Thiobenzaldehyde oxide (Phenylsulfine)

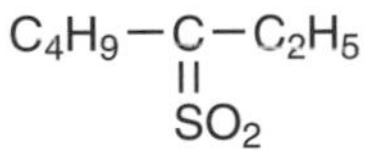

C$_4$H$_9-$C$-$C$_2$H$_5$ Heptane-3-thione dioxide (Ethyl butyl sulfene)
SO$_2$

Acetals and ketals as well as ortho esters can either be straightforwardly designated as derivatives of the underlying carbonyl compounds or named substitutively as **...oxy...** derivatives of the most senior parent structure actually present. Hemiacetals are treated as substituted alcohols.

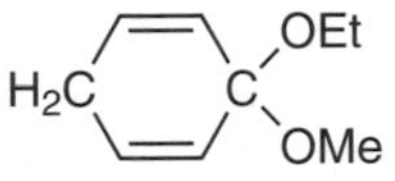

Cyclohexa-2,5-dien-1-one ethyl methyl ketal
or:
3-Ethoxy-3-methoxycyclohexa-1,4-diene

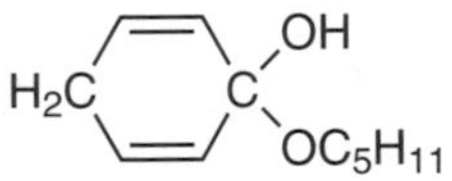

1-(Pentyloxy)cyclohexa-2,5-dien-1-ol

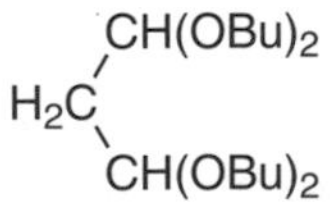

trad.:   Malonaldehyde tetrabutyl acetal
subst.: Propane-1,1,3,3-tetrayltetrakis(oxy)tetra-
         kisbutane (CA: ... 1,3-diylidentetrakis...)

Cyclic acetals can most simply be named as heterocycles. The

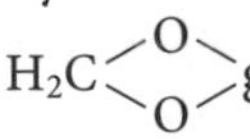

grouping is traditionally also treated as **methylenedioxy** substituent group.

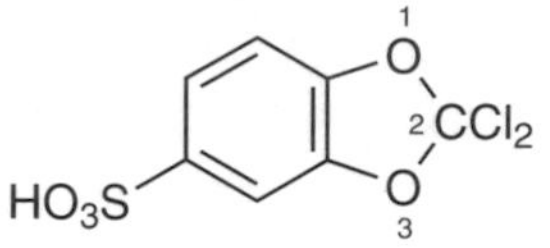

2,2-Dichloro-1,3-benzodioxole-5-sulfonic acid
or:
3,4-(Dichloromethylenedioxy)benzene-
sulfonic acid

Finally, **acylals** and **thioacetals** are henceforth only to be named substitutively as **esters** and **...ylthio...** compounds, respectively.

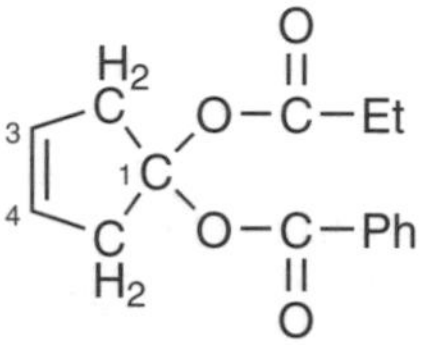

Cyclopent-3-enylidene benzoate propionate

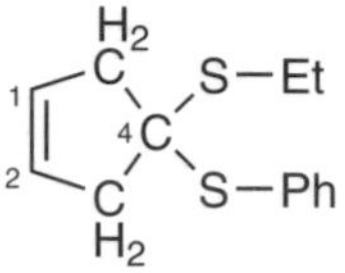

4-(Ethylthio)-4-(phenylthio)cyclopentene
(trad.: Cyclopent-3-en-1-one ethyl phenyl dithioketal)

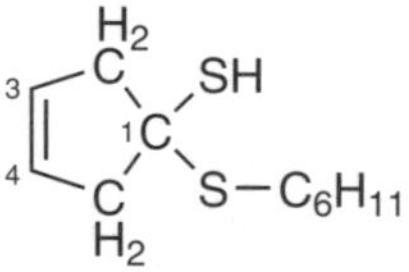

1-(Cyclohexylthio)cyclopent-3-ene-1-thiol

3.4
Alcohols, Phenols, and their Derivatives

Alcohols and phenols are generally named by the substitutive method although Chem. Abstr. applies conjunctive nomenclature whenever possible, i.e. for chain alcohols with ring substituent groups

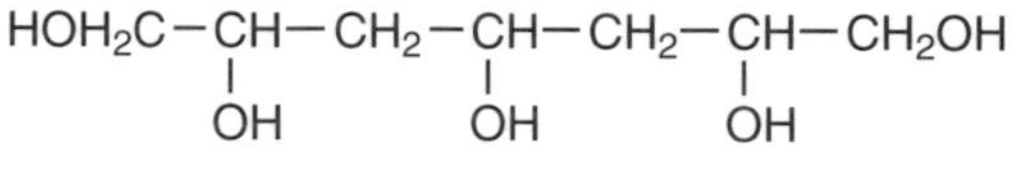

$HOH_2C-CH-CH_2-CH-CH_2-CH-CH_2OH$ Heptane-1,2,4,6,7-pentol
            | | |
            OH   OH   OH

$Cl_3C-CH(OH)_2$ $HC(OH)_3$

2,2,2-Trichloroethane-1,1-diol   Methanetriol
(Chloral hydrate)   (Orthoformic acid)

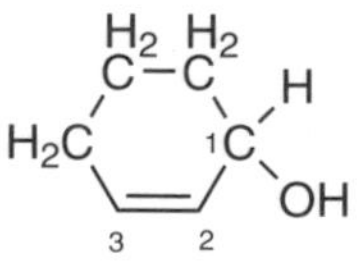

Cyclohex-2-en-1-ol

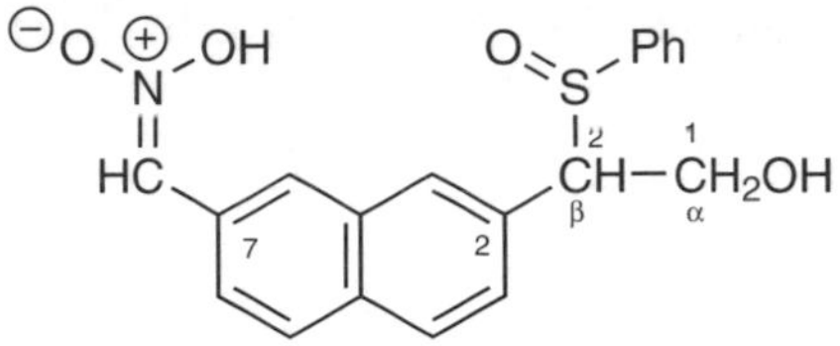

conj.: 7-*aci*-Nitromethyl-β-phenylsulfinylnaphthalen-2-ethanol

subst.: 2-(7-*aci*-Nitromethyl-2-naphthyl)-2-(phenylsulfinyl)ethanol

(According to a recent proposal **aci-nitro** should be replaced by **hydroxynitroryl**.)

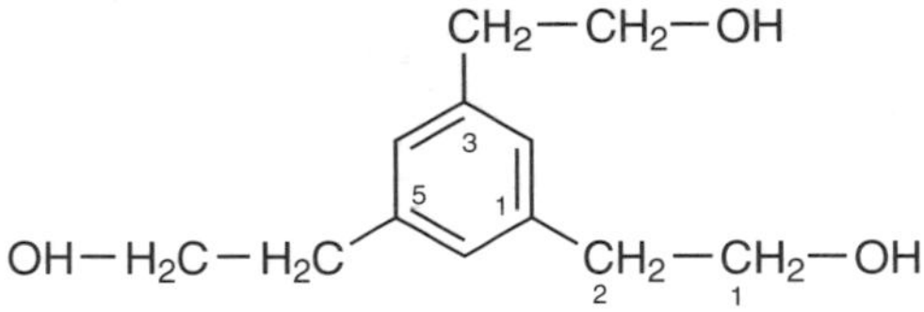

conj.: Benzene-1,3,5-triethanol

subst.: 2,2′,2″-(Benzene-1,3,5-triyl)trisethanol

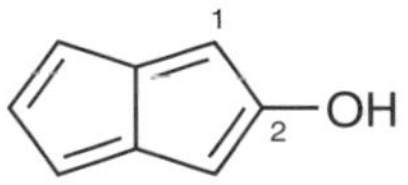

Pentalen-2-ol

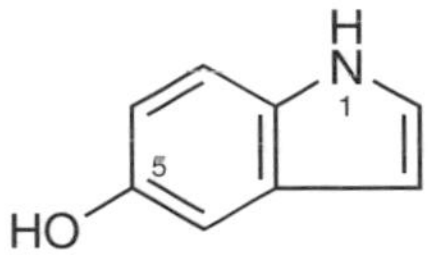

1*H*-indol-5-ol

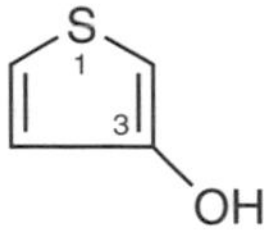

Thiophen-3-ol

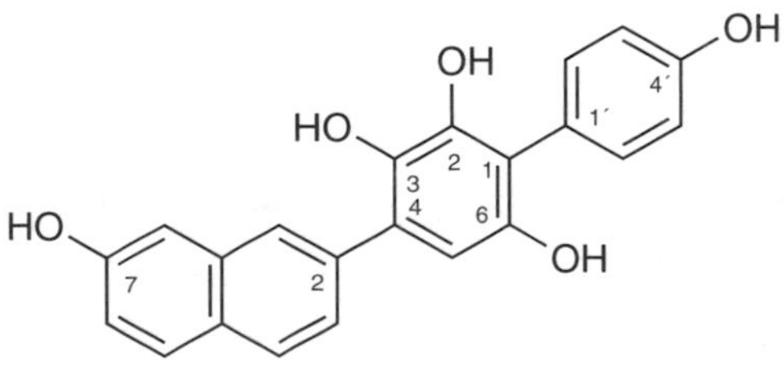

4-(7-Hydroxy-2-naphthyl)-biphenyl-2,3,6,4′-tetrol

Radicofunctional nomenclature is no longer used except for simple aliphatic and alicyclic alcohols, e.g.: methyl, pentyl, cyclohexyl alcohol, etc., and could be dispensed with here too.

Substituent groups derived from alcohols are characterized by the ending ... **yloxy.**

C_7H_{15}–O– Heptyloxy

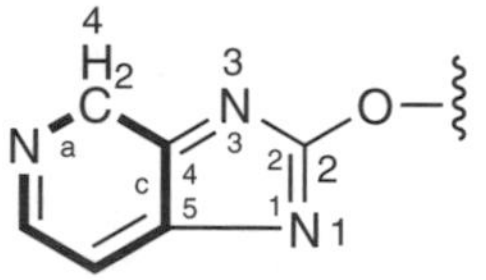

$4H$-[1,3]Imidazolo[4,5-c] pyridin-2-yloxy

The following abbreviated forms are retained: **methoxy, ethoxy, isopropoxy, (iso-, *sec*-, *tert*-)butoxy, phenoxy.** Divalent groups of that kind (–O–$(CH_2)_n$–O–) are designated as methylene-, ethylene-, trimethylenedioxy, etc. groups. Groups such as –O–CO–O–, –O–SO_2–O–, etc. are named carbonyldioxy, sulfonyldioxy etc. (see Section 2.2.6.2).

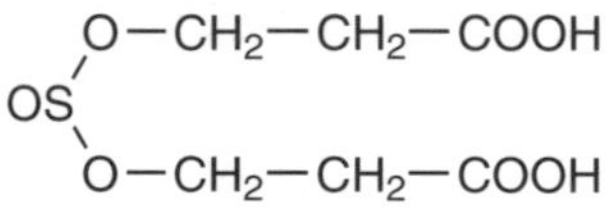

O–CH_2–CH_2–COOH

OS

O–CH_2–CH_2–COOH

3,3′-(Sulfinyldioxy)dipropanoic acid

Salts of alcohols and phenols whose names end with ... **ol** can generally be designated as ... **olates.**

(H_3C–CH_2–CH_2–O–$)_2$Mg   Magnesium di(propan-1-olate)

C_6H_5–CH_2–OLi

radicofunctional: Lithium benzylalcoholate
conjunctive: Lithium benzenemethanolate

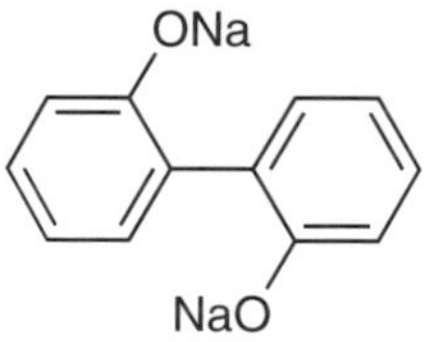

ONa

NaO

Disodium biphenyl-2,2′-di(olate)

Alternatively, **alcoholates** can also be denoted as ... **oxides** or, in a strictly systematic manner, as ... **oxidanides.**

(C_5H_{11}–O–$)_2$Ca   Calcium di(pentyloxide)
          or: Calcium di(pentyloxidanide)

H_3C–O–K   Potassium methoxide
          or: Potassium methyloxidanide

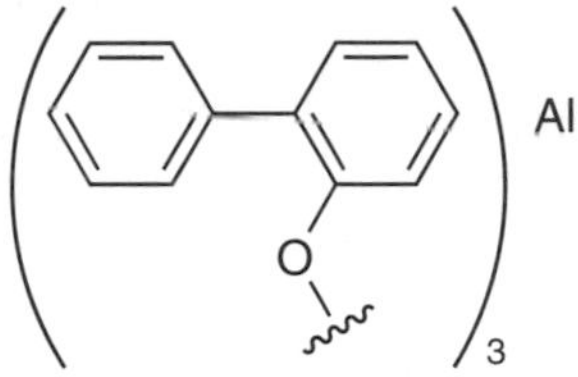

Al   Aluminumtri(biphenyl-2-yloxide)
      [new: Al tri(biphenyl-2-yloxidanide)]

Thioanalogs of alcohols are called **thiols;** the traditional names **mercaptane and mercapto group** (for –SH) should no longer be used. Here too, substitutive nomenclature is generally to be preferred. The corresponding salts are named following the procedures described for alcoholates.

$HS-CH_2-CH_2-CH_2-SH$ Propane-1,3-dithiol

C_6H_5-SH Benzenethiol (traditional name: thiophenol)

$C_6H_5-CH_2-SH$ Phenylmethanethiol
                         (Chem. Abstr.: Benzenemethanethiol)

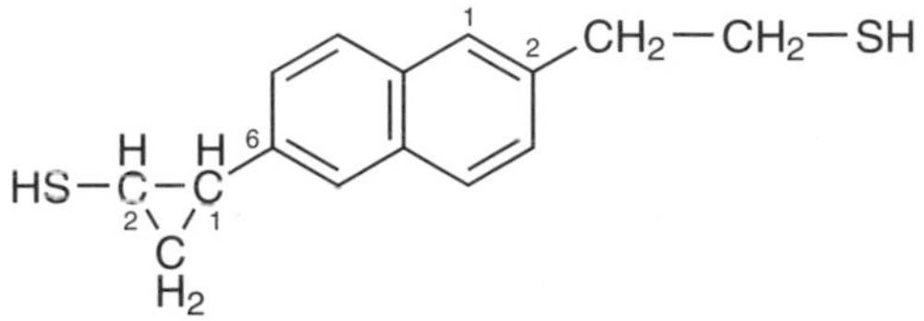

substitutively:
2-[6-(2-Sulfanylcyclopropyl)-2-naphthyl]ethanethiol
conjunctively:
6-(2-Sulfanylcyclopropyl)naphthalen-2-ethanethiol
radicofunctionally:
2-[6-(2-Sulfanylcyclopropyl)-2-naphthyl]ethyl hydrosulfide

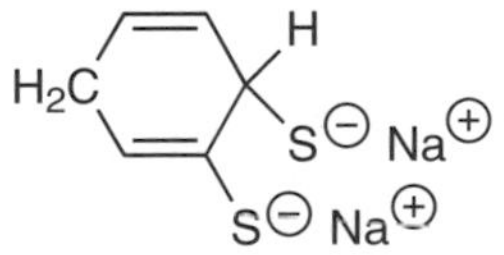

Disodium cyclohexa-2,5-diene-1,2-dithiolate

Monoorganyl derivatives of **chalcogen chains** are mostly still named radicofunctionally as **...yl hydropolychalcogenides** (exception: **...yl hydroperoxides**) although systematically they are to be perceived as **ylpolychalcogenanes.**

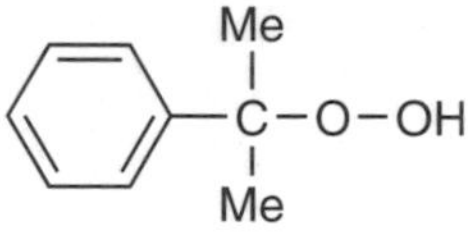

radicof.: α,α-Dimethylbenzyl hydroperoxide
substit.: (1-Phenyl-1-methylethyl)dioxidane

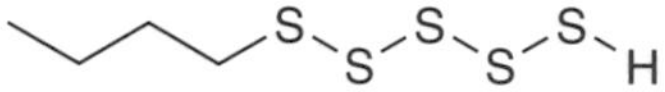

Butyl hydropentasulfide
Butylpentasulfane

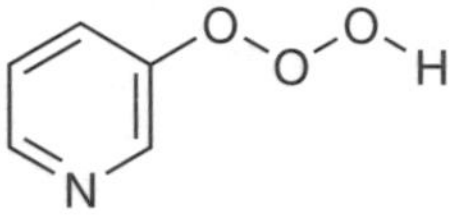

3-Pyridyl hydrotrioxide
Pyridin-3-yltrioxidane

3.5
Ethers and Thioethers

Ethers are named either **radicofunctionally** or **substitutively**, with a clear preference for the first method in simple cases.

radicof.: Diethyl ether
substit.: Ethoxyethane
CA: Oxybisethane

The previous example persuasively underscores how much simpler systematic nomenclature would be and above all how much easier to teach if general recognition of fully systematic names could eventually be reached for each and all parent hydrides (see Table 14); in the present case this would give the **substitutive** name **diethyloxidane**.

Cl–CH$_2$–CH$_2$–O–CH=CH$_2$ (2-Chloroethyl)vinyl ether
(2-Chloroethoxy)ethene

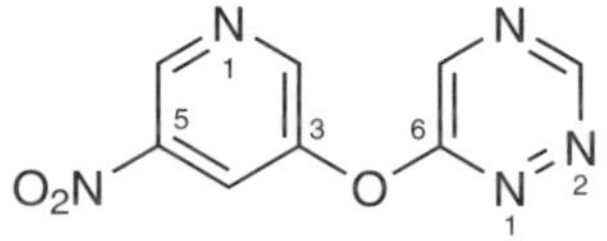

(1,2,4-Triazin-6-yl)(5-nitro-3-pyridyl) ether
6-(5-Nitro-3-pyridyloxy)-1,2,4-triazine

H$_3$C–O–CH$_2$–CH$_2$–O–CH$_3$ 1,2-Dimethoxyethane; **glycol dimethyl ether** is **not** a systematic name; radicofunctionally it would have to read: ethylene or ethane-1,2-diyl dimethyl ether!

If two identical functional parent structures are linked by an oxygen atom the nomenclature for **assemblies of identical units** is used as a rule.

HOOC-H$_2$C-CH$_2$-O-CH$_2$-CH$_2$-COOH   3,3'-Oxydipropionic acid

Cyclic ethers of the **epoxide** type can be named **additively** as ... **oxides, substitutively** as **epoxy** ... compounds (compare with bridged systems, p. 65) or, most uniformly, as **heterocycles.**

Me$-$C$-$CH$_2$ (with H above C and O below the C–CH$_2$)

Methyloxirane   (Propylene oxide is only correct
1,2-Epoxypropane if the trivial name **Propylene**
Propene oxide   instead of **Propene** is retained)

For linear polyethers and homologous compounds **replacement** nomenclature is recommended (p. 54).

Me$_{\diagdown}$O$\diagup$$\diagdown$S$\diagup$$\diagdown$S$\diagup$$\diagdown$$_2O\diagup$Me   2,8-Dioxa-4,6-dithianonane

Monoethers of polyvalent alcohols are best named substitutively but occasionally also quite trivially by attaching to the name of the alcohol the descriptive terms ... **monomethyl ether** or simply ... **methyl ether.**

CH$_2$$-O-C_3H_7$ 3-Propoxypropane-1,2-diol
CH$-$OH   or: Glycerol 1-propyl ether
CH$_2$$-$OH   or: 1-$O$-Propylglycerol

Individual species of the genus **thioether** can again most uniformly be named as ... **sulfane** and ... **sulfanyl** derivatives, respectively (formerly: ... **sulfides** and ... **thio** derivatives, respectively).

H$_3$C–S–CH$_2$–CH$_2$–CH$_2$–CH$_2$–CH$_2$–CH$_3$ Hexylmethylsulfane
                                    (or: ... sulfide)
                                    or: 1-(methylthio)hexane

H$_5$C$_2$–S–CH$_2$–S–CH$_2$–S–C$_2$H$_5$ Bis(ethylsulfanylmethyl)sulfane
                    formerly: Bis(ethylthiomethyl) sulfide

Cyclic sulfides (thioethers) are treated as **heterocycles,** in the same way as their ether counterparts. Polysulfides substituted at both ends are named substitutively as ... **polysulfanes** (formerly: ... **polysulfides**).

$H_3C-S-S-S-S-C_6H_5$ Methyl(phenyl)tetrasulfane
(Methyltetrasulfanyl)benzene

$Me-S-S-\langle\!\!\langle\;\;\rangle\!\!\rangle-S-S-S-S-S-COOMe$

4-[(Methyldisulfanyl)phenyl]pentasulfanecarboxylic acid methyl ester

When sulfides are oxidized at sulfur they transmute into two new compound classes traditionally designated as **sulfoxides** and **sulfones**, respectively. Accordingly, the most coherent method for naming individual members thereof is here exceptionally provided by **radicofunctional** nomenclature, since the practical application of substitutive nomenclature to these compounds as found in the registry nomenclature is to the highest degree incongruent and confusing.

On the one hand, relating simple sulfoxides to the most senior (but functionally subordinate) hydrocarbon component totally disregards the high rank of the SO and SO_2 functions. Polysulfoxides and polysulfones, on the other hand, continue to be named and indexed as such.

$$\begin{array}{c} O \\ \| \\ Me-\underset{\displaystyle ..}{S}-Me \end{array} \qquad\qquad \begin{array}{c} O \;\; O \\ \| \;\; \| \\ Me-\underset{\displaystyle ..}{S}-\underset{\displaystyle ..}{S}-Me \end{array}$$

radicof.:	Dimethyl sulfoxide	Dimethyl disulfoxide
IUPAC:	Methylsulfinylmethane	Dimethyl disulfoxide
CA:	Sulfinylbismethane	Dimethyl disulfoxide

C_4H_9-SO- [imidazo[4,5-d]thiazole ring structure]

Butyl-(4H-imidazo[4,5-d]thiazol-5-yl) sulfoxide
or 5-(Butylsulfinyl)-4H-imidazo [4,5-d]thiazole

$$H_3C-CH_2-CH\begin{array}{l} SO_2-C_5H_{11} \\[4pt] SO_2-C_5H_{11} \end{array}$$

radicof.: Propylidenebis(pentylsulfone)
substit.: 1,1'-[Propane-1,1-diylbis(sulfonyl)] bispentane

In this context it may be again noted that rigorous application of the existing rules of substitutive nomenclature on the basis of the universal **parent hydride names** of Table 14 would lead to solutions of utmost consistency and simplicity, e.g.:

$$\begin{array}{c} O \;\; O \\ ^3\| \;\; \|^2 \;\; ^1 \\ Me-\underset{\displaystyle ..}{S}-S-S-Me \\ \| \\ O \end{array}$$

Dimethyl-1,2λ^6,3λ^4-trisulfane-2,2,3-trione
(Dimethyl-1,2,3-trisulfane 2,2,3-trioxide)

Related imino derivatives are characterized as ... **sulfimides,** ... **sulfox-imides;** (Chem. Abstr. uses here names ending with ... **sulfimines** and ... **sulfoximines,** respectively).

N-Bromo-S,S-dimethyl sulfoximide

Cyclic sulfoxides and sulfones ar best reated additively as **oxides** and **diox-ides,** respectively (see p. 87).

3.6
Amines, Imines, and their Derivatives

For amines too **radicofunctional, conjunctive,** and **substitutive** naming methods are so confusingly practiced side by side that a unifying treatment has long been overdue. This would most easily be accomplished by adopt-ing the systematic parent hydride name **azane;** as second choice, though, at least those substitutive names should be exclusively used which result from attaching the suffix ... **amine** to the unchanged name of the parent. By the way, the latter approach would be concordant with the pertinent indexing modes of Chem. Abstr.

H_3C-NH_2
substit.: Methanamine
new: Methylazane
radicof.: Methyl amine

Naphthalen-1-amine
Naphthalen-1-ylazane
1-Naphthylamine

substit.: 9aH-Quinolizin-3-amine
radicof.: 9aH-Quinolizin-3-yl amine
new: (9aH-Quinolizin-3-yl)azane

An acyclic amine terminally substituted by a cyclic system is named either **substitutively** or **conjunctively; (radicofunctional** and **azane** names may be obtained as described above).

subst.: 3-(9bH-Phenalen-2-yl)propanamine
conj.: 9bH-Phenalene-2-propanamine

Primary **diamines** and **polyamines** are most consistently named according to **substitutive** principles.

$H_2N-CH_2-CH_2-CH_2-CH_2-NH_2$ Butane-1,4-diamine

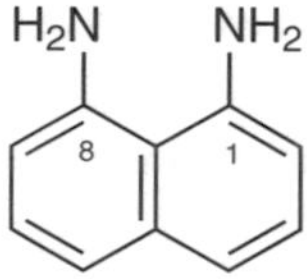

Naphthalene-1,8-diamine

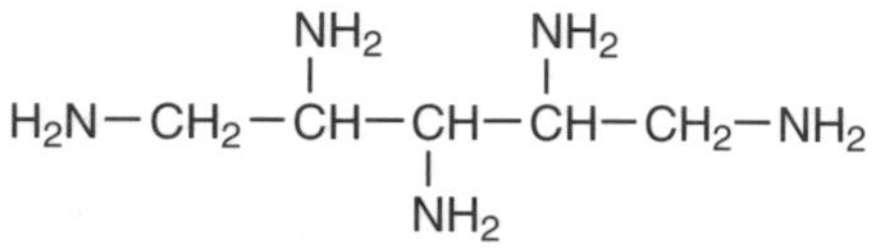

Pentane-1,2,3,4,5-pentamine

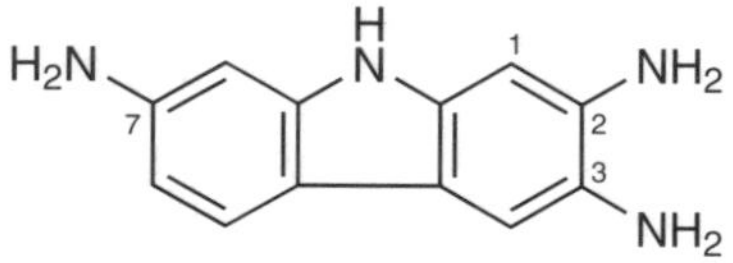

Carbazole-2,3,7-triamine

If all amino groups are found in the **side chains** of a cyclic system, **substitutive** or **conjunctive** names are utilized.

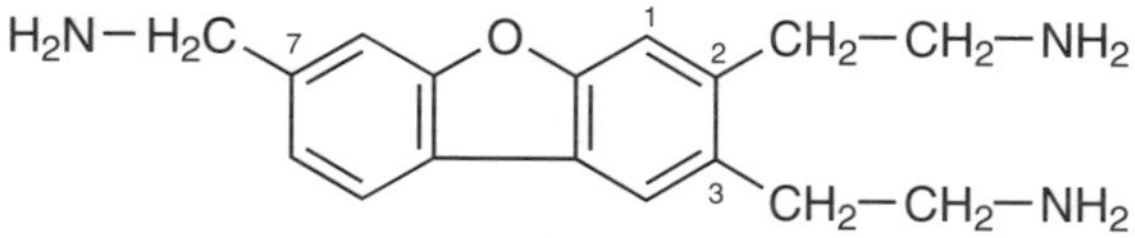

conj.: 7-(Aminomethyl)dibenzofuran-2,3-bis(ethanamine)
subst.: 2,2′-[7-(Aminomethyl)dibenzofuran-2,3-diyl]bis(ethanamine)

Secondary and **tertiary amines** are in most cases still named **radicofunctionally,** in a manner that can easily be transformed into a substitutive naming mode based on the parent hydride name **azane.** Within the boundaries of conventional substitutive nomenclature, amines of that sort are interpreted as products of *N*-**substitution** of the most serior primary amine present.

$(C_6H_5)_3N$ Triphenyl-amine; (new: Triphenyl azane)
N,N-Diphenyl-benzenamine

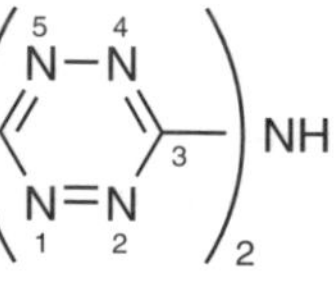

Di-(1,2,4,5-tetrazin-3-yl)amine (new: ... azane)
N-(1,2,4,5-Tetrazin-3-yl)-1,2,4,5-tetrazin-3-amine

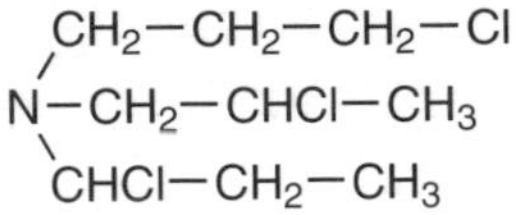

$$CH_2-CH_2-CH_2-Cl$$
$$N-CH_2-CHCl-CH_3$$
$$CHCl-CH_2-CH_3$$

(1-Chloropropyl)-(2-chloropropyl)-(3-chloro-propyl)amine
(new: ... azane)
1-Chloro-N-(2-chloropropyl)-N-(3-chloro-propyl)propanamine

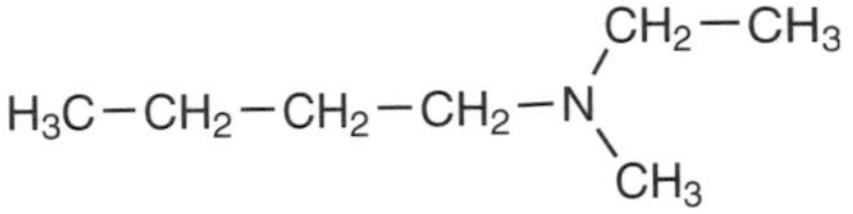

$$H_3C-CH_2-CH_2-CH_2-N\begin{smallmatrix}CH_2-CH_3\\CH_3\end{smallmatrix}$$

(Butyl) (ethyl) (methyl)amine
(...azane)
N-Ethyl-N-methylbutan-1-amine

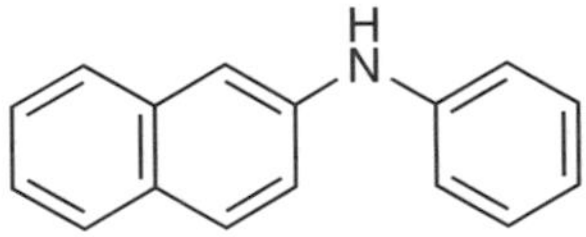

(2-Naphthyl) (phenyl)amine (... azane)
N-Phenylnaphthalen-2-amine

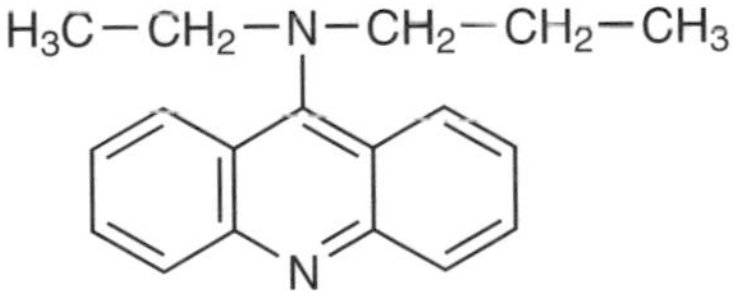

$$H_3C-CH_2-N-CH_2-CH_2-CH_3$$

(Acridin-9-yl) (ethyl) (propyl)amine
(... azane)
N-Ethyl-N-propylacridin-9-amine

Acylated amines are essentially treated as acid amides of the most senior acid present.

$(C_5H_9-CO-)_3N$ N,N-Bis(cyclopentylcarbonyl)cyclopentane-carboxamide

$(C_6H_5-CO-)_2NH$ N-Benzoylbenzamide

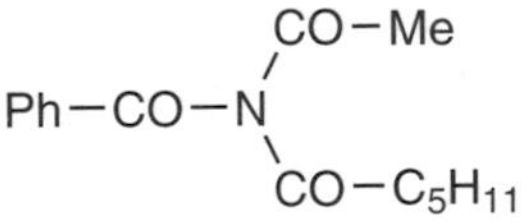

$$Ph-CO-N\begin{smallmatrix}CO-Me\\CO-C_5H_{11}\end{smallmatrix}$$

N-Acetyl-N-hexanoylbenzamide

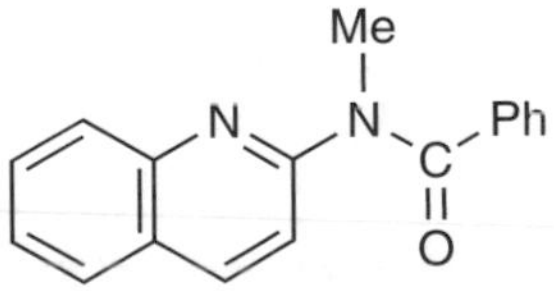

N-(Quinolin-2-yl)-N-methylbenzamide

Phenyl-substituted acid amides are commonly designated as **anilides**.

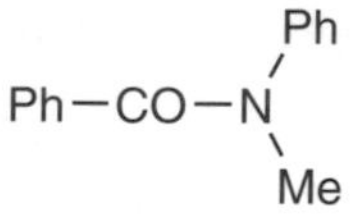

Ph—CO—N $\begin{array}{c}\text{Ph}\\ \diagup \\ \diagdown \\ \text{Me}\end{array}$ N-Methylbenzanilide
(N-Methyl-N-phenylbenzamide)

Amides whose nitrogen atom is part of a ring are generally regarded as **acylated heterocycles,** subordinate groups gaining preference only when attached to the same ring.

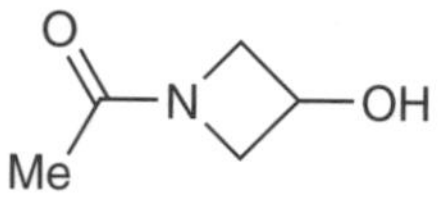

1-Acetylazetidin-3-ol

but:

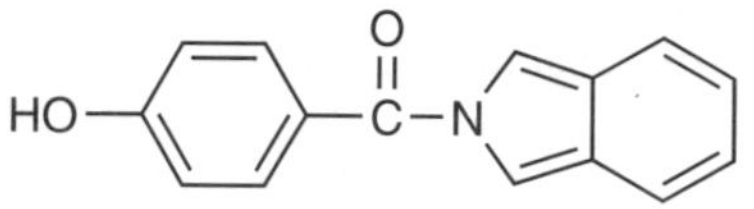

2-(4-Hydroxybenzoyl)-2H-indole

For **lactams** and similar compounds in most cases the conventional trivial names of Table 8 are applied although here too heterocyclic nomenclature would render much more uniform names possible.

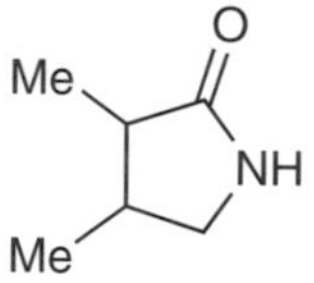

3,4-Dimethylpyrrolidin-2-one

Imines form the functional compound class of lowest rank and are hence named as such only when no N-organyl substituents are present; otherwise they are named as **...ylidenamines.** If more senior functions are to be respected **...imino** prefixes are used. Totally systematic **azane**-based names would again bring about considerable simplifications.

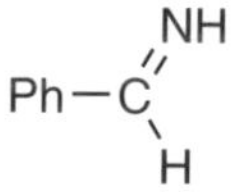

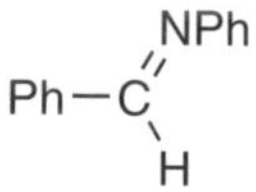

Ph—C $\begin{array}{c}\text{NH}\\ \diagup \\ \diagdown \\ \text{H}\end{array}$ Ph—C $\begin{array}{c}\text{NPh}\\ \diagup \\ \diagdown \\ \text{H}\end{array}$

subst.: Phenylmethanimine   syst.: N-(Phenylmethylene)benzenamine
conj.: Benzenemethanimine   trad.: Benzyliden- or Benzalaniline
new: (Phenylmethylen)azane   new: Phenyl(phenylmethylen)azane

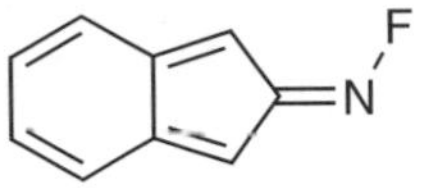

N-Fluoro-2H-indenimine
[Fluoro-(2H-inden-2-
yliden)azane]

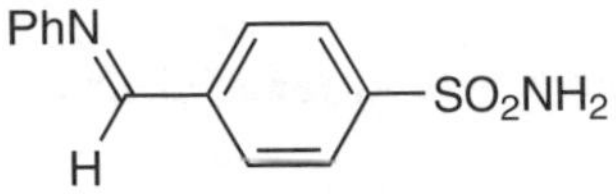

4-(Phenyliminomethyl)benzene-
sulfonamide [4-(Phenylazanylidene-
methyl)...]

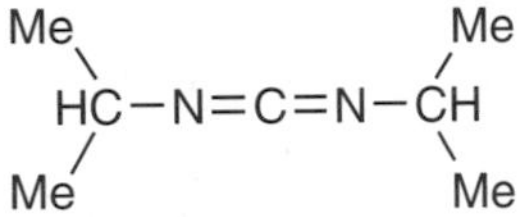

$$Me-\overset{Me}{\underset{Me}{HC}}-N=C=N-\overset{Me}{\underset{Me}{CH}}$$

syst.: N,N'-Methanedi(ylidene)bis(propan-2-
amine)
CA.: ...methanetetraylbis...
trad.: N,N'-Diisopropylcarbodiimide
new: Methanedi(ylidene)bis(propan-2-
ylazane)

3.7
Halogen Derivatives

Halogen compounds should generally be named substitutively; complete substitution can be indicated by the prefix **per**.

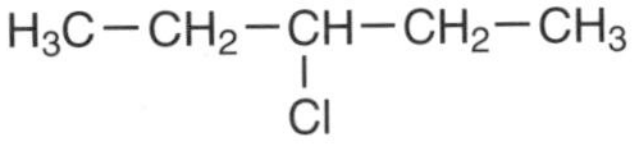

$$H_3C-CH_2-\overset{}{\underset{Cl}{CH}}-CH_2-CH_3$$

3-Chloropentane

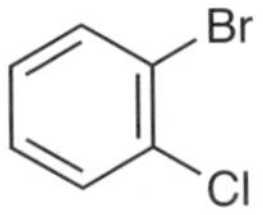

1-Bromo-2-chlorobenzene
(o-Bromochlorobenzene)

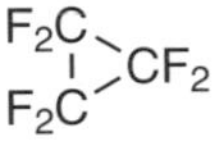

$$\overset{F_2C}{\underset{F_2C}{>}}CF_2$$ Perfluorocyclopropane

For some simple halogen derivatives, however, **radicofunctional** names are still widely in use.

H_3C-I	Methyl iodide (iodomethane)
BrH_2C-CH_2Br	Ethylene dibromide (1,2-dibromoethane)
$C_6H_5-CHCl_2$	Benzylidene dichloride, traditionally also benzal chloride

In a few cases additive names still persist although they are strongly discouraged by now.

$C_6H_5-CHBr-CHBr-C_6H_5$ Stilbene dibromide (1,2-dibromo-1,2-
diphenylethane)
Chem. Abstr.: 1,2-Dibromoethane-1,2-
diylbisbenzene

3.8
Compounds with Nitrogen Chains

3.8.1
Azo and Azoxy Compounds

Although a thoroughly consistent, systematic approach to the naming of virtually all **azo** and **azoxy compounds** – compounds which have traditionally been treated very disparately – on the basis of the highly versatile parent hydride name **diazene** has existed for some time, several customary older naming modes are still perpetuated in pretty confusing ways (even by Chem. Abstr.!). These demand a more detailed discussion.

Undisputed are **diazene** names for derivatives containing hydrocarbon, alkoxy, and comparable substituent groups or directly attached senior functionalities.

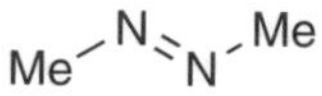

Dimethyldiazene
(trad.: Azomethane)

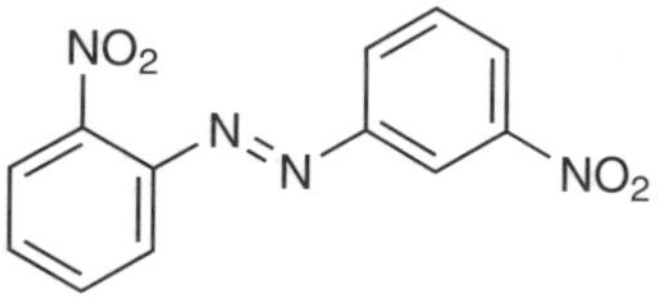

(2-Nitrophenyl)(3-nitrophenyl)diazene
(trad.: 2,3'-Dinitroazobenzene)

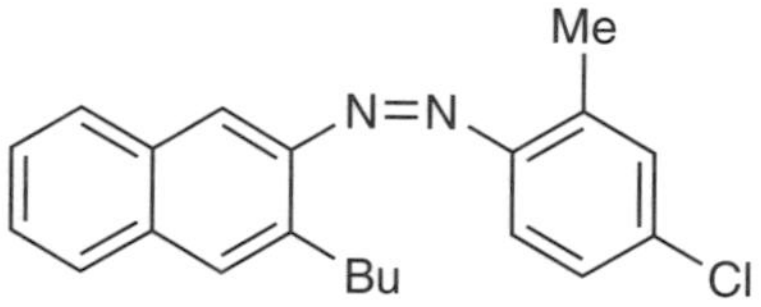

syst.: (3-Butylnaphthalen-2-yl)-(4-chloro-2-methylphenyl)diazene
formerly: 2-Butyl-3-(4-chloro-2-methylphenylazo)naphthalene
trad.: 3-Butylnaphthalen-2-azo-(4'chloro-2'-methylbenzene)

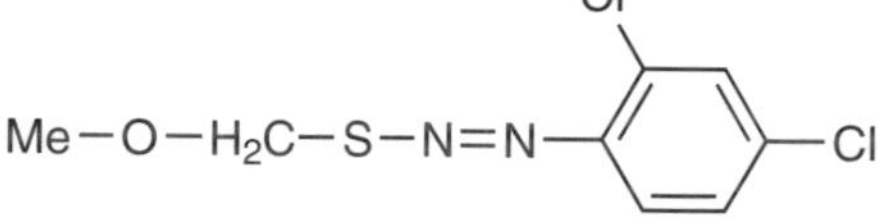

(2,4-Dichlorophenyl)(methoxymethylsulfanyl)diazene

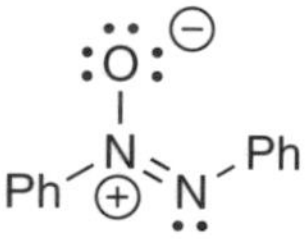

Diphenyldiazene oxide
(Azoxybenzene)

1-[(3-Methylphenylsulfonyl)oxy]-
2-phenyldiazene 2-oxide

N,N-Dimethylthiophen-2-yldiazene-
carboxamide

MeO_3S–N=N–SO_3Me

Diazenedisulfonic acid dimethyl ester or
Dimethyl diazenedisulfonate
(Azodisulfonic acid dimethyl ester or
Dimethyl azodisulfonate)

First discrepencies are then already discernible for the covalent isomers of
arenediazonium salts, which are properly named as **aryldiazenols** by
IUPAC but, contrary to the rules, as **arylhydroxydiazanes** by Chem. Abstr.

C_6H_5–N=N–OH

Systematic name: Phenyldiazenol
Chem. Abstr. Hydroxy(phenyl)diazene
Traditional name: Benzenediazo-
hydroxide

C_6H_5–N=N$\ddot{\underset{..}{O}}$:$^{\ominus}$ Na$^{\oplus}$

Sodium phenyldiazenolate
Hydroxy(phenyl)diazene,
sodium salt
Sodium benzenediazoate

Yet another treatment is operative in the naming of doubly alkoxy-substituted
diazenes which are denoted as **esters of hyponitrous acid** by Chem. Abstr.

MeO–N=N–OMe   syst.: Dimethoxydiazene
CA: Hyponitrous acid dimethyl ester

Unfortunately the situation becomes worse still: even if one can come to
terms with the decision to name azo derivatives of nitrogen cycles as
diazenyl-substituted heterocycles – because of the plausible preference of
rings over chains – it remains utterly incomprehensible why the **unsubstituted diazenyl group** is named as such, whereas its **substituted equivalents**
are identified as ... **ylazo** groups.

4-Diazenylpyridine

syst.: 4-(Phenyldiazenyl)pyridine
CA: 4-(Phenylazo)pyridine

In the presence of higher ranked functional groups too, Chem. Abstr. uses exclusively the traditional ... *azo* ... infixes.

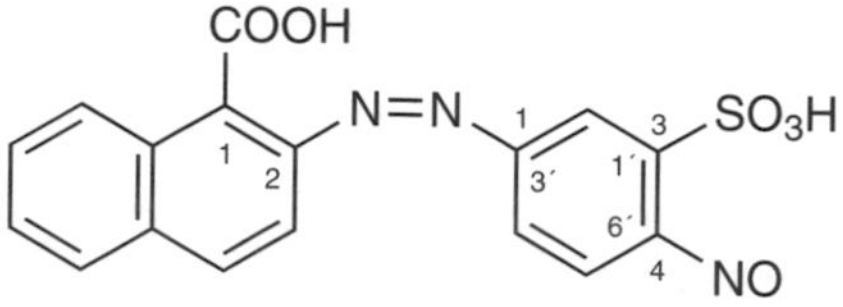

syst.: 2-[(4-Nitroso-3-sulfophenyl)diazenyl]naphthalene-1-carboxylic acid
CA: 2-(4-Nitroso-3-sulfophenylazo)naphthalene-1-carboxylic acid
trad.: 1-Carboxynaphthalene-2-azo-3'-(6'-nitrosobenzenesulfonic acid)

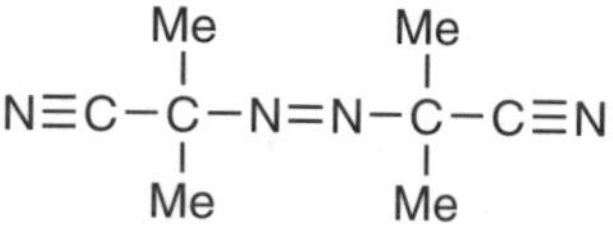

syst.: 2,2'-Dimethyl-2,2'-diazene-diylbis(propanenitrile)
CA: 2,2'-Azobis(2-methylpropanenitrile)
trad.: Azoisobutyronitrile

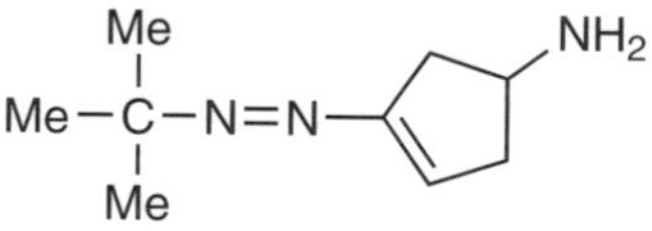

syst.: 3-[(1,1-Dimethylethyl)dia-zenyl]cyclopent-3-en-1-amine
CA: 3-[(1,1-Dimethylethyl)azo] ...

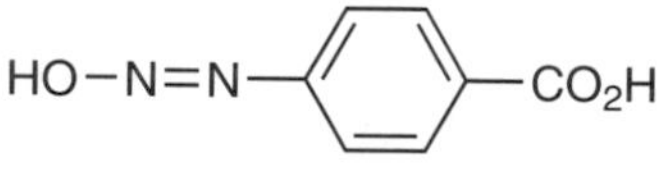

syst.: 4-(Hydroxydiazenyl)benzoic acid
CA: 4-(Hydroxyazo)benzoic acid

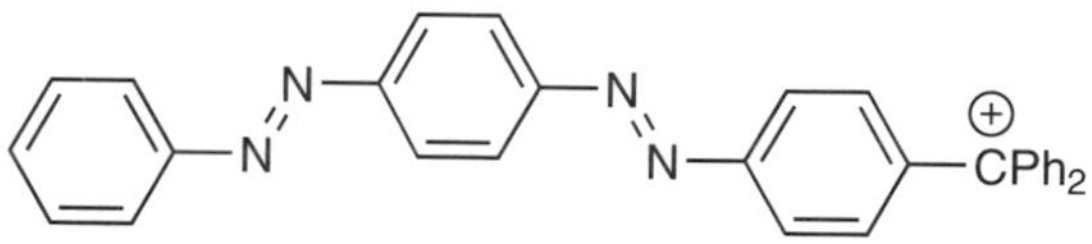

Diphenyl-{4-[4-(phenyldiazenyl)phenyldiazenyl]phenyl}methylium

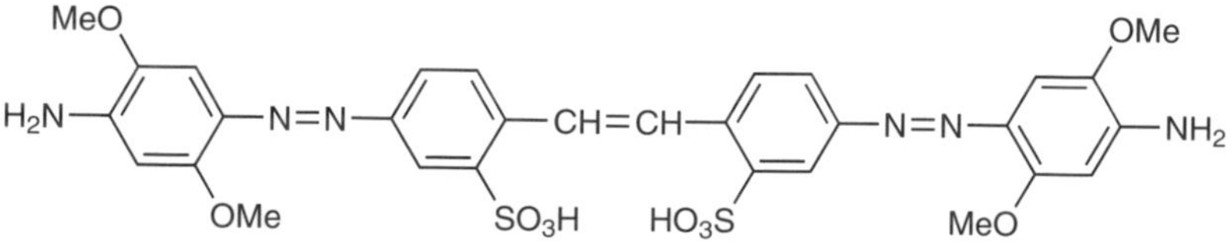

2,2'-(Ethene-1,2-diyl)bis-{5-[(4-amino-2,5-dimethoxyphenyl)dia-zenyl]benzolsulfonic acid}

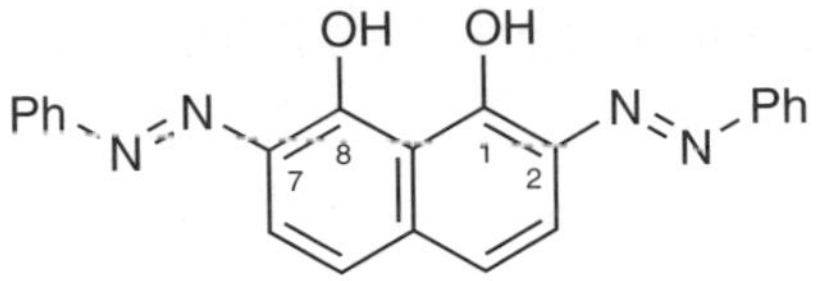

2,7-Bis(phenyldiazenyl)naphtalene-1,8-diol

Symmetrical bis- and **poly-azo compounds** with peripheral senior functions are treated as assemblies of identical components (see Section 2.2.6).

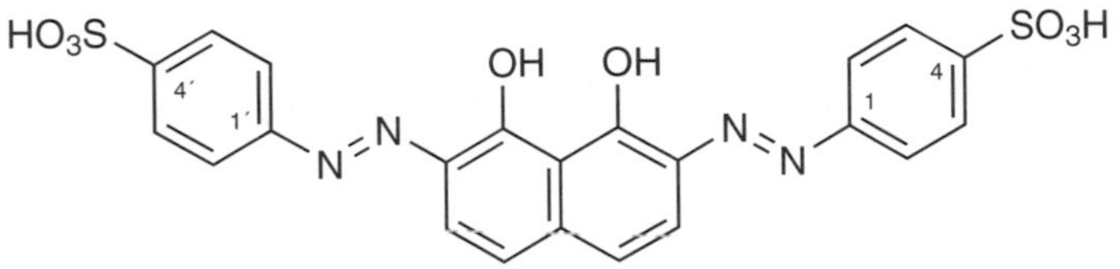

syst.: 4,4'-[(1,8-Dihydroxynaphthalene-2,7-diyl)bis(diazendediyl)]bisbenzenesulfonic acid
CA: 4,4'-(Dihydroxynaphthalene-2,7-diylbisazo)bisbenzenesulfonic acid

If **azoxy functions** must be expressed as prefixes even IUPAC abandons the path of systematic virtue and insists on the traditional ... *ONN*... and ... *NNO*... infixes although in these cases too complete systematization is feasible.

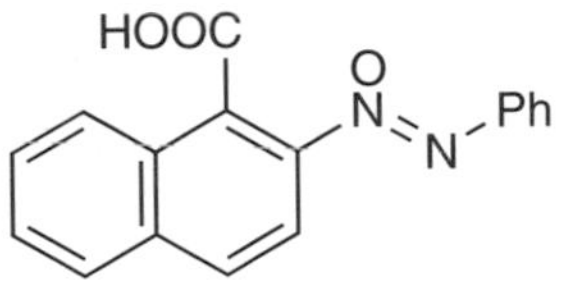

trad.: 1-Carboxynaphthalene-2-*ONN*-azoxybenzene
IUPAC/CAS: 2-(Phenyl-*NNO*-azoxy)naphthalene-1-carboxylic acid
syst.: 2-(1-Oxo-2-phenyl-1λ^5-diazenyl)naphthalene-1-carboxylic acid

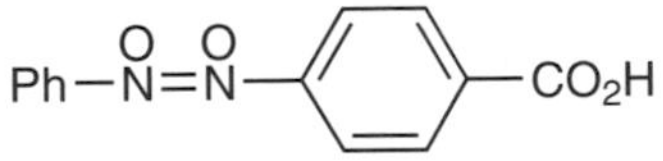

4-(1,2-Dioxo-2-phenyl-1λ^5,2λ^5-diazenyl) benzoic acid

3.9
Hydrazines and their Derivatives

Simple substitution products of hydrazines (with the exception of acyl-substituted ones) are generally still named as such and, where necessary, in the form of ... **hydrazino**... prefixes, although here again the change to systematic **diazane** names would greatly enhance the uniformity of the pertinent naming procedures.

$C_6H_5-NH-NH_2$ Phenylhydrazine (Phenyldiazane)

$C_6H_5-NH-NH-C_6H_5$ 1,2-Diphenylhydrazine (...diazane)
Trivial name: Hydrazobenzene

$(H_3C)_2N-N(C_4H_9)_3$ N,N-Dibutyl-N',N'-dimethylhydrazine
(1,1-Dibutyl-2,2-dimethyldiazane)

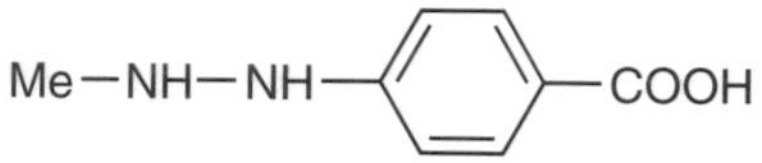

4-(N'-Methylhydrazino)benzoic acid
4-(2-Methyldiazanyl)benzoic acid

Acyl-substituted hydrazines are classified as acid hydrazides and named by transforming the ending ...ic acid into ...**ohydrazide**. Chem. Abstr. applies here again paraphrases such as ...**ic acid, hydrazides,** ...**ylhydrazide.**

$H_3C-CH_2-CO-NH-NH-CH_2-CH_3$ N'-Ethylpropionohydrazide
(1-Ethyl-2-propanoyldiazane)

Chem. Abstr.:   Propanoic acid, hydrazides,
2-ethylhydrazide

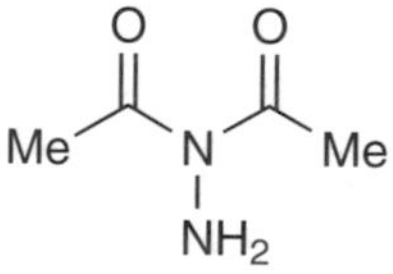

N,N-Diacetylhydrazine
(1,1-Diacetyldiazane)
CA: Acetic acid, hydrazides, 1-acetyl hydrazide

3.10
Diazo and Diazonium Compounds

Diazonium salts and **aliphatic diazo compounds** can easily be named following the instructions layed down in Tables 6 and 8.

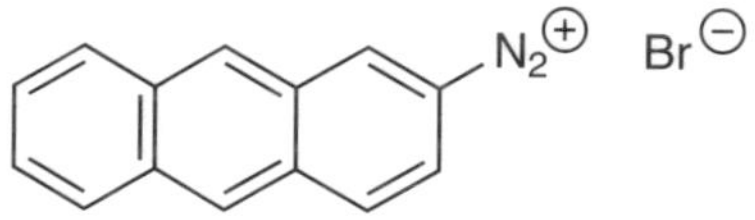

Anthracene-2-diazoniumbromide

H_3C-CHN_2 Diazoethane $C_6H_5-CO-CHN_2$ 2-Diazo-1-phenyl-
ethanone
Trivial name:
Diazoacetophenone

Compounds of type RN=NX having X linked directly to the diazo group have already been treated in Section 3.8.

3.11
Compounds with Chains of Three or More Nitrogen Atoms

The first compounds to be considered here are **azides,** which are mostly named **radicofunctionally** as ...**yl azides** or **substitutively** as **azido**... derivatives.

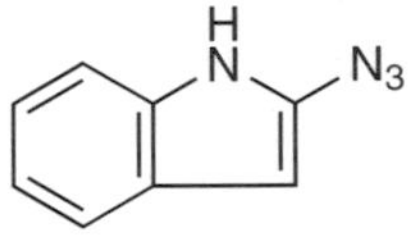

1*H*-Indol-2-yl azide or 2-Azido-1*H*-indole

$C_6H_5-CO-N_3$ Benzoyl azide

All other compounds containing such nitrogen chains are designated as **polyazanes, -azenes, -azadienes,** etc. with double bonds being preferred for numbering.

$H_2N-NH-NH-CH_3$ 1-Methyltriazane

$H_3C-NH-N=N-C_6H_5$ 3-Methyl-1-phenyltriazene

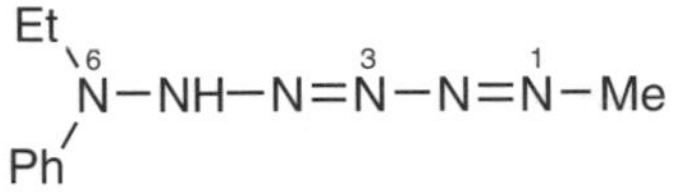

6-Ethyl-1-methyl-6-phenylhexaaza-1,3-diene

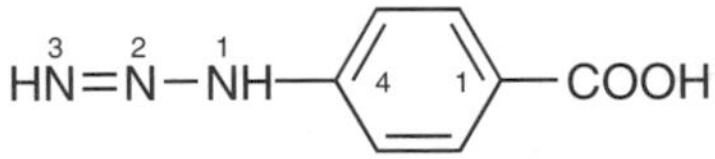

4-(Triaz-2-enyl)benzoic acid

Triazenes substituted at both ends with identical organyl groups were formerly denoted as **diazoamino**... compounds.

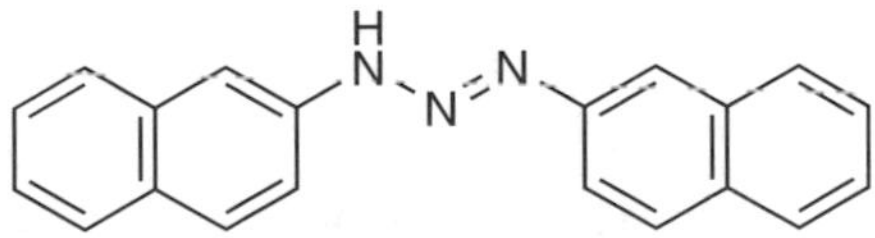

2,2'-Diazoaminonaphthalene
[1,3-Di(naphthalen-2-yl)triazene]

3.12
Other Polynitrogen Parent Systems

Trivial names of freely substitutable **polynitrogen** parent structures as recommended for further use by IUPAC and the most important substituent groups derived therefrom are compiled in Table 13. Differing names and numberings as employed by Chem. Abstr. are enclosed in parentheses.

Table 13. Retained trivial names for polynitrogen parent structures[a]. (In parentheses: differing Chem. Abstr. names and numberings.)

$$\underset{4\quad 3\quad 2\quad 1}{H_2N-\overset{\overset{O}{\|}}{C}-NH-CO_2H}$$

Allophanic acid
(Aminocarbonylcarbamic acid)

$$\underset{4\qquad\qquad 1}{H_2N-\overset{\overset{O}{\|}}{C}-NH-\overset{\overset{O}{\|}}{C}-\xi}$$

Allophan(o)yl
[(Aminocarbonyl)aminocarbonyl]

$$\underset{5(N')\ (3)\ \ 3(2)\ \ (1)\ \ 1(N)}{H_2N-\overset{\overset{(N''')4}{\overset{NH}{\|}}}{C}-NH-\overset{\overset{NH^{2(N'')}}{\|}}{C}-NH_2}$$

Biguanide
(Imidodicarbonimidic diamide)

$$\underset{5(N')\quad 4\quad\ 3\quad\ 2\quad 1(N)}{H_2N-\overset{\overset{O}{\|}}{C}-NH-\overset{\overset{O}{\|}}{C}-NH_2}$$

Biuret[b]
(Imidodicarbonic diamide)

$$\underset{5(1)\ \ 4(2)\ \ 3\ \ 2(4)\ \ 1(5)}{HN=N-\overset{\overset{O}{\|}}{C}-NH-NH_2}$$

Carbazone
(Diazenecarboxylic acid, hydrazide)

$$\underset{5\qquad\qquad\qquad 1}{HN=N-\overset{\overset{O}{\|}}{C}-NH-NH-\xi}$$

Carbazono
(Diazenylcarbonylhydrazino)

$$\underset{5\ \ 4\ \ 3\ \ 2\ \ 1}{HN=N-\overset{\overset{O}{\|}}{C}-N=NH}$$

Carbodiazone

$$HN=C=NH$$

Carbodiimide
(Methanediimine)

$$\underset{5(2')\ \ 4(1')\ \ 3\ \ 2(1)\ \ 1(2)}{H_2N-NH-\overset{\overset{O}{\|}}{C}-NH-NH_2}$$

Carbonohydrazide
(Carbonic dihydrazide)

$$\underset{5(1)\ \ 4(2)\ \ 3\ \ 2(4)\ \ 1(5)}{HN=N-CH=N-NH_2}$$

Formazan

$$\underset{(1)\qquad\qquad (5)}{\xi-N=N-CH=N-NH_2}$$

(1-Formazano)

$$\underset{(1)\qquad\qquad\ (5)}{HN=N-CH=N-NH-\xi}$$

(5-Formazano)

Table 13 (continued)

$HN = N - C = N - NH_2$

(Formazanyl)

$-N = N - C = N - N =$
(1)   (3)   (5)
(Formazan-1,3-diyl-5-ylidene)

$2(N'') NH$
$\qquad \parallel$
$H_2N - C - NH_2$
3(N')   1(N)

Guanidine[c]

NH
$\parallel$
$H_2N - C - NH -$

Guanidino
(Aminoiminomethyl)amino

O
$\parallel$
$H_2N - C - NH_2$
3(N')   2   1(N)

Urea

O
$\parallel$
$H_2N - C - NH -$
3   2   1

Ureido
[(Aminocarbonyl)amino]

OH
$\vert$
$HN = C - NH_2$
3   2   1

Isourea
(Carbamimidic acid)

O
$\parallel$
$H_2N - C - NH - CH_2 - CO_2H$
5   4   3   2   1

Hydantoic acid
[N-(Aminocarbonyl)glycine]

O   O
$\parallel$ $\parallel$
$H_2N - C - NH - CH_2 - C -$
5   1

Hydantoyl
[N-(Aminocarbonyl)glycyl]

O
$\parallel$
$H_2N - C - NH - NH_2$
4   3   2   1

Semicarbazide
(Hydrazinecarboxamide)

O
$\parallel$
$H_2N - C - NH - NH -$
4   3   2   1

Semicarbazido
[2-(Aminocarbonyl)hydrazino]

O
$\parallel$
$H_2N - C - NH - N =$

Semicarbazono
((Aminocarbonyl)hydrazono)

---

[a] Chalcogen analogues: thiourea, selenocarbazide, tellurocarbodiazone etc.
[b] Analogously: Triuret etc.
[c] Analogously: Biguanide, triguanide etc.

4 Metalorganic and Metalloidorganic Compounds

On account of the plethora of organometall(oid)ic compounds and the ever contentious question of their assignment to organic chemistry on the one hand or inorganic chemistry on the other, the development of a binding systematic nomenclature for this domain always was and still is fraught with major difficulties and is thus far from being settled. All that can be given here, therefore, is just a survey of current practice supplemented by sketches of a future scenario determined by emerging tendencies towards tighter systematization. Meanwhile it has become abundantly clear that at least two naming variants will ultimately persist: one organic, substitutive, and one inorganic, coordinative.

4.1
Element Hydride (Elementane) Names

The substitutive naming variant is customarily applicable in all cases where weakly polar, molecular, i.e. non-aggregated compounds with (at least formally) 2-center 2-electron bonds are concerned. On the basis of the element hydride concept, according to which the names of these fundamental structures containing only hydrogen and the respective element are formed through combination of appropriatly modified "a" terms from Tables 4, 20 with the ending ...ane, carbanes (hydrocarbons) are thus closely associated with **heteranes** (element hydrides) in a way that leaves little choice but to adopt the unifying collective term **elementane**.

Regrettably this simple and above all easily explicable treatment has so far been sanctioned by **IUPAC** only for hydrides of element groups 13–17 (with the exception of aluminum – thallium) (Table 14), although it can be applied in equally self-evident manner to all other elements of the periodic table, e.g.: λ^6-wolframane (tungstane) (WH$_6$), λ^4-titanane (TiH$_4$), λ^5-niobane (NbH$_5$), λ^2-magnesane (MgH$_2$) etc.; possibly even: λ^7-ununseptane for [117] H$_7$!

Organic derivatives of these fundamental hydrides are then to be named as **organylelementanes.**

Table 14. Systematic names for parent hydrides (elementanes) of element groups 13–17

13		14[b]		15[b,c]		16[b]		17[b]	
BH_3	Borane	CH_4	Carbane	NH_3	Azane	OH_2	Oxidane	FH	Fluorane
AlH_3	Aluminane	SiH_4	Silane	PH_3	Phosphane	SH_2	Sulfane	ClH	Chlorane
GaH_3	Gallane	GeH_4	Germane	AsH_3	Arsane	SeH_2	Selane	BrH	Bromane
InH_3	Indi(c)ane	SnH_4	Stannane	SbH_3	Stibane	TeH_2	Tellane	IH	Iodane
TlH_3	Thallane	PbH_4	Plumbane	BiH_3	Bismutane	PoH_2	Polane	AtH	Astatane

[a] The possibilitiy to assign systematic names to all of these fundamental hydrides should be no means prejudice the abolishment of the perennial trivial designations **methane, ammonia, water,** and **hydrogen halides.**

[b] For hydrides with non-standard valencies of the core atom the following ruling holds: $EH_n = \lambda^n$-**elementane.**

[c] For the elements of group 15 Chem. Abstr. still uses the traditional designations phosph-, ars-, stib-, bismuth**ine.**

$(C_6H_{11})_2PH$	Dicyclohexylphosphane
$[(CH_3)_3C]_3B$	Tri-t-butylborane
$(CF_3C_6H_4)_6Te$	Hexakis-(4-trifluormethylphenyl)-λ^6-tellane
$(CH_3)_4S$	Tetramethyl-λ^4-sulfane
$[(CH_3)_3C-CH_2]_3I$	Trineopentyl-λ^3-iodane
$(i\text{-Pr})_4Ti$	Tetraisopropyl-λ^4-titanane
$(CF_3)_2Hg$	Bis(trifluormethyl)mercurane
Me_6W	Hexamethyl-λ^6-wolframane (tungstane)
Ph-Pd-H	Phenyl-λ^2-palladane

Tetrakisacetyloxy-(1,1′:3′,1″-terphenyl-2′-yl)-λ^5-iodane

$(EtO)_3SiH$	Triethoxysilane
$(C_4H_9)InH_2$	Butylindi(c)ane

When applying **inorganic coordination nomenclature** to such compounds they can be named essentially without any prior supposition by prefixing the name of the ligand group (including hydrogen = hydrido) to the unchanged element term of the central atom.

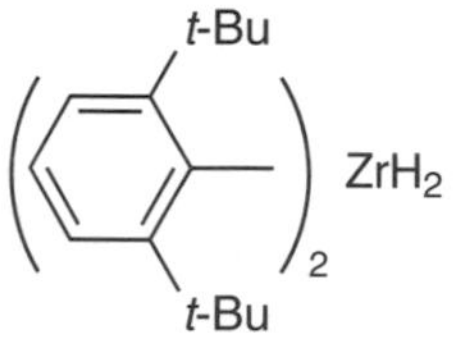

ZrH$_2$ Bis-(2,6-di-t-butyl)dihydridozirconium

(CF$_3$)$_5$Nb   Pentakis(trifluoromethyl)niobium

Ph$_3$Br   Triphenylbromine

t-BuO-AlH(t-Bu) t-Butoxy-t-butylhydridoaluminum

When naming **mixed element hydrides** and their organic derivatives it would be logical per se to follow the priority order laid down in the complete list of replacement terms (Table 20) accepted by inorganic and organic chemists alike. This, however, conflicts with the ranking selected by Chem. Abstr. for its substantially longer (compared to IUPAC) priority list of compound classes, where heteranes of element groups 14–16 plus boron take precedence over hydrocarbons of any kind. This leads to the following totally non-systematic priority sequence:

P > As > Sb > Bi > B > Si > Ge > Sn > Pb > O > S > Se > Te > C

Cl$_2$HC-PH-SiH$_2$Cl   (Chlorosilyl)(dichloromethyl)phosphane

Me$_3$Se-SbPh$_4$ Tetraphenyl(trimethyl-λ^4-selanyl)-λ^5-stibane

F$_3$S≡C-SF$_3$ Trifluoro(trifluoro-λ^4-sulfanylmethylidyne)-λ^6-sulfane

t-Bu-B[Si(i-Pr)$_3$]$_2$ t-Butylbis(triisopropylsilyl)borane

4.2
Functionally Substituted Elementanes

If such compounds actually contain the typical **functional groups** of organic chemistry, the **customary ranking order** of **functional compound classes** (see Table 7) within the limits of **substitutive** or **conjunctive** (Chem. Abstr.) nomenclature is reinstated, then requiring all organometall(oid)ic fragments to be named as substituent prefixes.

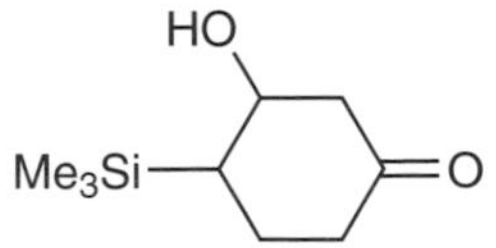

3-Hydroxy-4-trimethylsilylcyclohexanone

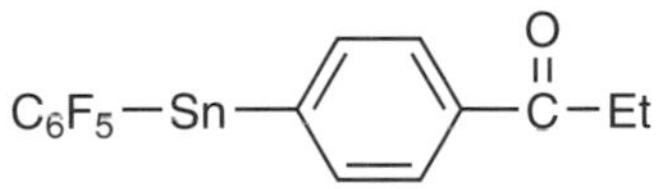

1-[4-(Pentafluorophenyl-λ^2-stannyl)-phenyl]propan-1-one

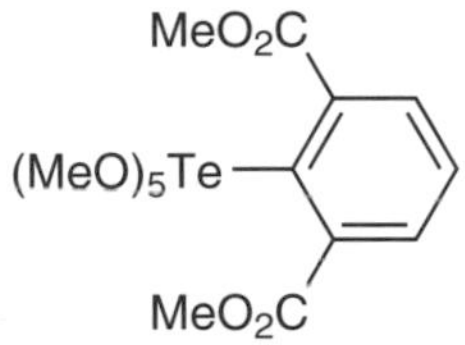

(Pentamethoxy-λ^6-tellanyl)benzene-1,3-dicarboxylic acid dimethyl ester

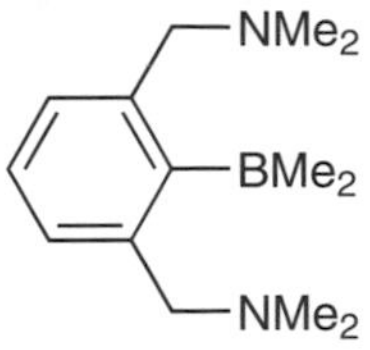

substitutively: 2-Dimethylboryl-*N,N,N′,N′*-tetramethylbenzene-1,3-diylbismethanamine
conjunctively (Chemical Abstracts):
2-Dimethylboryl-*N,N,N′,N′*-tetramethylbenzene-1,3-dimethanamine

Reflecting the significance of coordination nomenclature, names focusing on the metal(loid) atom are also frequently used here: e.g. dimethyl-[2,6-bis(dimethylaminomethyl)phenyl]boron for the last example above.

4.3
Elementanes with Repeating Diads

Special rules are valid for chains and rings assembled from **regularly ordered diads** of two **different element types.** Chains here must end with identical atoms appearing as late as possible in the "a" term Tables 4 and 20, rings must be made up of identical duplexes (repeating or repetitive units). Compound names are then formed by combining the respective "a" terms in the reverse order to that shown in Tables 4 and 20 and by prefixing them with numerical terms indicating the number of terminal atoms in the case of chains and the number of identical units for rings.

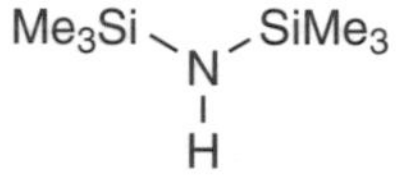

$Me_3Si \diagdown N \diagup SiMe_3$
    |
    H

1,1,1,3,3,3-Hexamethyldisilazane

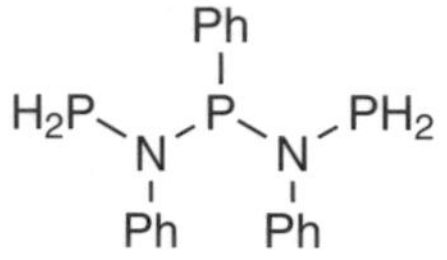

$H_2P \diagdown N \diagup P \diagdown N \diagup PH_2$
                    |
        Ph   Ph
(with Ph on the central P)

2,3,4-Triphenyltriphosphazane

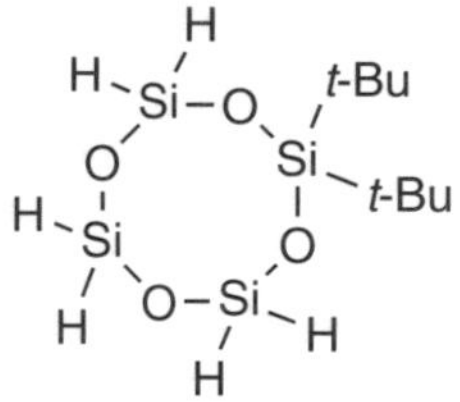

2,2-Di-t-butylcyclotetrasiloxane

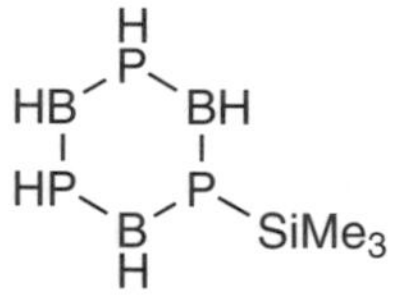

1-Trimethylsilylcyclotriboraphosphane

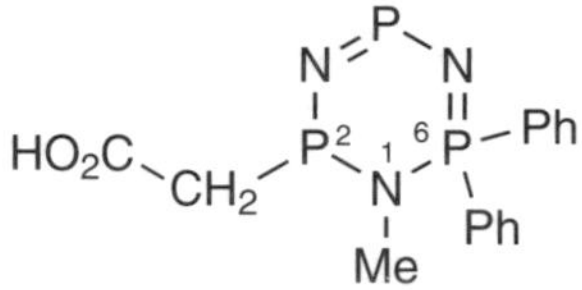

(1-Methyl-6,6-diphenyl-6λ^5-cyclotriphos-phaza-3,5-dien-2-yl)acetic acid
or conjunctively: ... 3,5-dien-2-acetic acid

4.4
Organic Derivatives of Alkali and Alkaline Earth Metals and Comparable Compounds

In contrast to all preceding names conventional names for the highly polar alkali and alkaline earth metal-organic compounds only rarely reflect the factual connectivities between metal atom and organic group(s) since in reality these compound types comprise predominantly **oligomeric aggregates** and/or **solvated complexes**. In a simplifying manner, however, their names are usually formed from the fundamental monomeric structural units by a) prefixing the name of the organic substituent group to the unchanged name of the metal – thus following inorganic **coordination nomenclature** – or b) regarding these substances as **salts** of **organic anions.**

C_4H_9Li a) Butyllithium
                 b) Lithium butanide

$(C_6H_5)_3CNa$ a) Triphenylmethylsodium
                 (trivially: tritylsodium)
                 b) Sodium triphenylmethanide

$(H_3C)_3CLi$ a) t-Butyllithium
                 b) Lithium 2-methylpropan-2-ide

$H_2C=CHLi$ a) Ethenyllithium
                 (Trivial name: vinyllithium)
                 b) Lithium ethenide

a) Biphenyl-2,2'-diyldilithium
b) Dilithium biphenyl-2,2'-diide
(or also: 2,2'-Dilithiobiphenyl, see below)

a) 2-Butylpyridin-1(2H)-yllithium
= 2-Butyl-1,2-dihydropyridin-1-yllithium
b) Lithium 2-butylpyridin-1(2H)-ide
= Lithium 2-butyl-1,2-dihydropyridin-1-ide

a) 1-Methyl-1-phenylethylpotassium
= 2-Phenylpropan-2-ylpotassium
(trivially: α,α-Dimethylbenzylpotassium)
b) Potassium 2-phenylpropan-2-ide

$H_3C\equiv CRb$ a) Prop-1-ynylrubidium
                 b) Rubidium propyn-1-ide

$Cs^{\oplus}$ a) Cyclopentadienylcaesium
b) Caesium cyclopentadienide

For the less polar organic derivatives of the alkaline earth metals (with the exception of Grignard compounds) method a) is applied exclusively.

$(C_6F_{5})_2Mg$ Bis(pentafluorophenyl)magnesium

$(Ph_3C)_2Ca$ Bis(triphenylmethyl)calcium

Me_2Ba Dimethylbarium

Metalorganyls of the Grignard compound type, however, are predominantly named as hemi salts according to a hybrid procedure based on method a) – if coordination names are not preferred outright.

C_6H_5MgI Phenylmagnesium iodide
Iodo(phenyl)magnesium

H_3CMgH Methylmagnesium hydride
Hydridomethylmagnesium

$(H_3C)_3CSrBr$ t-Butylstrontium bromide
t-Butylbromostrontium

$(C_6H_5)_3CCaCl$ Triphenylmethylcalcium chloride
Chloro(triphenylmethyl)calcium

It is immediately apparent that analogous organic derivatives of transition metals can be named in exactly the same manner:

$C_6H_5-C{\equiv}C-Ag$ Phenylethynylsilver

(2,4,6-Tris-t-butylphenyl)gold dichloride
Dichloro(tri-t-butylphenyl)gold
why not:
Dichloro(tri-t-butylphenyl)-λ^3-aurane

ICH_2ZnI	Iodomethylzinc iodide Iodomethyliodozinc
$(CH_3)_2Cd$	Dimethylcadmium
C_6H_5HgOAc	Phenylmercury acetate Acetato(phenyl)mercury
$(F_3C)_2Hg$	Bis(trifluormethyl)mercury
$(C_3H_7)_2TiCl_2$	Dichlorodipropyltitanium Dipropyltitanium dichloride
$[(H_3C)_2HC]_2Zr[OCH(CH_3)_2]_2$	Diisopropylzirkonium diisopropoxide Diisopropyldiisopropanolatozirkonium

For multiply metalated parent hydrides in particular another naming method borrowed from inorganic nomenclature can also be applied; hereby metal substituents are expressed in the form of **metalio** prefixes, e.g.:

1,3,5-Trilithiobenzene

Li_6C	Hexalithio-λ^6-methane (or: Hexalithio-λ^6-carbane) or simply: Hexalithiocarbon
$(EtHg)_4Si$	Tetrakis(ethylmercurio)silane or Tetrakis(ethylmercurio)silicon
$[(Ph_3PAu)_5Sb]^{2\oplus}$	Pentakis(triphenylphosphanaurio)stibanedi(ium)

4.5
"ate" Complexes

Beyond the broad spectrum of sometimes only vaguely distinguishable structural types encountered among metalorganic species and having been dealt with, if only cursorily, in the preceding paragraphs, all these compounds can in principle combine to discrete salt-like aggregates of **"ate" complex** nature which can be named as such, at least in those cases where the definite anion structure is known for sure.

$[(Me_3Si)_3C-\overset{\ominus}{Na}-C(SiMe_3)_3]Na^{\oplus}S_n$ Sodium bis(trimethylsilyl-
                                      methyl)natrate (sodiate)
                                      (S = solvent)

$[(H_3C)_3CCH_2Mg^{\oplus}]\{\overset{\ominus}{Mg}[CH_2C(CH_3)_3]\}$ Neopentylmagnesium
                                      trineopentyl magnesate

$Li^{\oplus}[\overset{\ominus}{In}(C_4H_9)_3]$ Lithium tributylindi(c)ate

$Li^{\oplus}[\overset{\ominus}{Cu}(CH_3)_2]$ Lithium dimethylcuprate

$Li_2^{\oplus}[\overset{2\ominus}{Fe}(C_3H_7)_4]$ Dilithium tetrapropylferrate

$Na^{\oplus}[\overset{\ominus}{B}(C_6F_5)_4]$ Natrium tetrakis(pentafluoro-
                                      phenyl)borate

$Li^{\oplus}[\overset{\ominus}{Bi}(C_6H_5)_6]$ Lithium hexaphenylbismuthate

5 Carbohydrates

It goes without saying that by strictly applying the rules of subtitutive nomenclature, all carbohydrates and their derivatives could readily and uniformly be named as **polyhydroxyalkanals, -alkanones, -tetrahydrofurans, -tetrahydropyrans, -oxepanes, -alkanoic acids,** etc. which would, however, require supplementing by a more or less extended set of stereodescriptors. It is precisely this last problem which in the specific framework of carbohydrate nomenclature has been solved in a distinctly different and certainly much clearer manner, in that numerous stereochemically unambiguously defined trivial and semitrivial names have been installed as cornerstones of the body of rules. The pertinent rule manual has recently been published in a thoroughly revised edition but its fundamental systematic features, as presented in the following paragraphs, should also be familiar to the non-specialist.

5.1
Aldoses

Monosaccharides of the polyhydroxyalkanal type have the class name **aldose** and as individual representatives with up to six carbon atoms the trivial names compiled in Table 15.

Table 15. Trivial names[a] for aldoses with recommended three-letter symbols

CHO H—C—OH CH$_2$OH	CHO H—C—OH H—C—OH CH$_2$OH	CHO HO—C—H H—C—OH CH$_2$OH	
D-Glyceraldehyde D-*glycero*	D-Erythrose D-*erythro*	D-Threose D-*threo*	
CHO H—C—OH H—C—OH H—C—OH CH$_2$OH	CHO HO—C—H H—C—OH H—C—OH CH$_2$OH	CHO H—C—OH HO—C—H H—C—OH CH$_2$OH	CHO HO—C—H HO—C—H H—C—OH CH$_2$OH
D-Ribose D-*ribo* (D-Rib)	D-Arabinose D-*arabino* (D-Ara)	D-Xylose D-*xylo* (D-Xyl)	D-Lyxose D-*lyxo* (D-Lyx)
CHO H—C—OH H—C—OH H—C—OH H—C—OH CH$_2$OH	CHO HO—C—H H—C—OH H—C—OH H—C—OH CH$_2$OH	CHO H—C—OH HO—C—H H—C—OH H—C—OH CH$_2$OH	CHO HO—C—H HO—C—H H—C—OH H—C—OH CH$_2$OH
D-Allose D-*allo* (D-All)	D-Altrose D-*altro* (D-Alt)	D-Glucose D-*gluco* (D-Glc)	D-Mannose D-*manno* (D-Man)
CHO H—C—OH H—C—OH HO—C—H H—C—OH CH$_2$OH	CHO HO—C—H H—C—OH HO—C—H H—C—OH CH$_2$OH	CHO H—C—OH HO—C—H HO—C—H H—C—OH CH$_2$OH	CHO HO—C—H HO—C—H HO—C—H H—C—OH CH$_2$OH
D-Gulose D-*gulo* (D-Gul)	D-Idose D-*ido* (D-Ido)	D-Galactose D-*galacto* (D-Gal)	D-Talose D-*talo* (D-Tal)

[a] The affiliation with the D- or L-series is determined by the configuration of the highest numbered stereogenic center, the *configurational atom*. For the customary Fischer projections the following conventions apply: HO-pointing to the right = D, to the left = L; horizontal bonds are directed towards the viewer while vertical bonds are in the plane or are orientated away from the viewer.

Cyclic hemiacetal forms derived from these fundamental types are systematically named as ...ooxiroses (3-ring), ...ooxetoses (4-ring), ...ofuranoses (5-ring), ...opyranoses (6-ring), ...oseptanoses (7-ring), e.g.:

β-L-Glucooxetose

β-D-Glucofuranose

α-D-Glucopyranose

α-L-Glucoseptanose

The configurational symbols α, β refer each to the so-called **anomeric reference atom** that in the above examples is identical with the **configurational atom**. Accordingly, the *cis*-relationship between the HO group attached to the anomeric C atom and the oxygen attached to the reference atom translates into α and *trans* into β.

In reality, the cyclic hemiacetal forms of sugars and their derivatives may assume a wealth of different ring conformations. For purposes of nomenclature, however, linear **Fischer Projections** and **Haworth Representations** with their quasi conformation-averaging planar-cyclic habit are best suited to illustrate the manifold stereochemical relationships significant for carbohydrates.

For the fundamental sugar types of Table 15 the respective trivial names are generally preferred to the systematic names obtained according to the pattern D-*ribo*-pentose (for D-ribose), D-*gluco*-hexose (for D-glucose), etc. For higher aldoses, however, exclusively systematic names are formed with **configurational prefixes** derived from Table 15 and **stem terms** such as ...**heptose**, ...**octose** etc., encoding chain length. The configurational prefixes are assigned according to the configurational patterns of the group of four stereogenic centers following C-1, ignoring interposed non-stereogenic atoms.

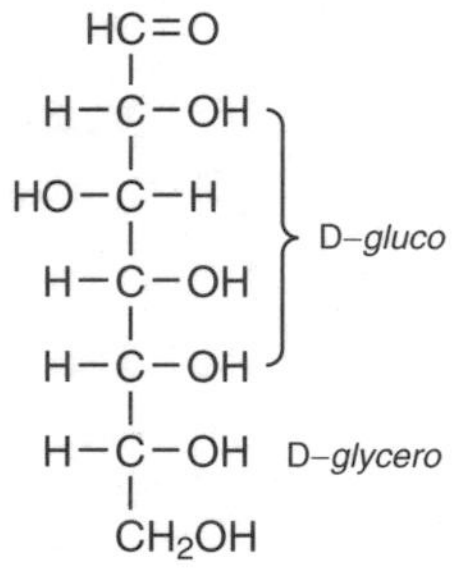

D-*glycero*-D-
gluco-Heptose

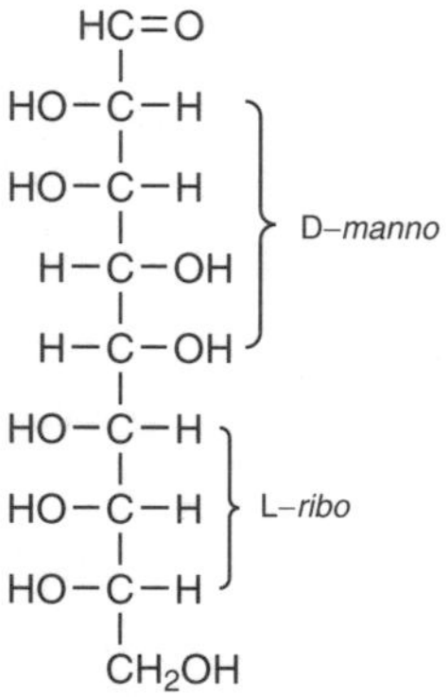

L-*ribo*-D-*manno*-
Nonose

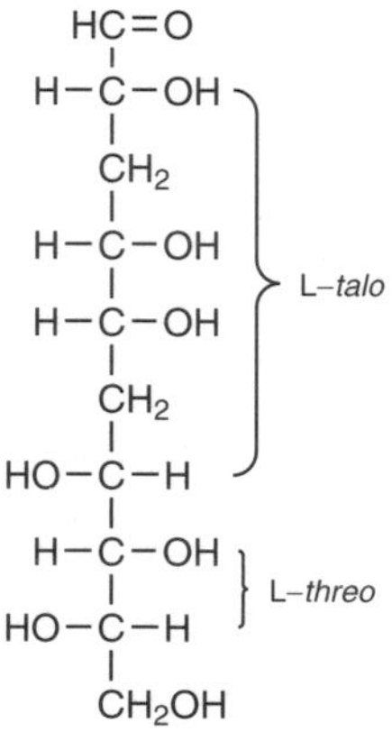

3,6-Didesoxy-L-*threo*-
L-*talo*-Decose

Systematic names for dialdoses are constructed similarly, e.g.: L-*threo*-tetrodialdose, D-*gluco*-hexodialdose, etc.

5.2
Ketoses

In the case of ketoses too trivial names are still frequently preferred for the fundamental types although systematic ...**ulose** names can easily be generated in a manner similar to that used for the systematic names of aldoses (see Table 16).

Systematic names for higher ketoses are generated in the same way as those for higher aldoses by combining stem terms such as ...**hept-n-ulose**, ...**oct-n-ulose**, etc. with the appropriate configurational prefixes taken from Table 15 and again ignoring keto groups in position 3 and higher.

Table 16. Trivial names with three-letter-short forms and, in round brackets, systematic names for ketoses

```
CH2OH              CH2OH              CH2OH              CH2OH
|                  |                  |                  |
C=O                C=O                C=O                C=O
|                  |                  |                  |
CH2OH           H−C−OH            H−C−OH            HO−C−H
                   |                  |                  |
                   CH2OH           H−C−OH            H−C−OH
                                      |                  |
                                      CH2OH              CH2OH
```

| 1,3-Dihydroxy-acetone | D-Erythrulose (D-*glycero*-Tetrulose) | D-Ribulose; D-Rul (D-*erythro*-Pent-2-ulose) | D-Xylulose; D-Xul (D-*threo*-Pent-2-ulose) |

```
CH2OH              CH2OH              CH2OH              CH2OH
|                  |                  |                  |
C=O                C=O                C=O                C=O
|                  |                  |                  |
H−C−OH          HO−C−H            H−C−OH            HO−C−H
|                  |                  |                  |
H−C−OH          H−C−OH            HO−C−H            HO−C−H
|                  |                  |                  |
H−C−OH          H−C−OH            H−C−OH            H−C−OH
|                  |                  |                  |
CH2OH              CH2OH              CH2OH              CH2OH
```

| D-Psicose; D-Psi (D-*ribo*-Hex-2-ulose) | D-Fructose; D-Fru (D-*arabino*-Hex-2-ulose) | D-Sorbose; D-Sor (D-*xylo*-Hex-2-ulose) | D-Tagatose; D-Tag (D-*lyxo*-Hex-2-ulose) |

```
        CH2OH                           CH2OH
        |                               |
        C=O                          H−C−OH ⎤
        |                               |   |
     HO−C−H                           C=O   |
        |         ⎤                     |    ⎬ D–allo
     H−C−OH       |                 H−C−OH  |
        |          ⎬ D–altro           |    |
     H−C−OH       |                 H−C−OH  |
        |         |                     |   ⎦
     H−C−OH       ⎦                 H−C−OH ⎤
        |                               |   ⎬ L–threo
        CH2OH                        HO−C−H ⎦
                                        |
                                        CH2OH
```

D-*altro*-Hept-2-ulose   L-*threo*-D-*allo*-Non-3-ulose
(Sedoheptulose)

Systematic names for **diketoses** are generated analogously, as usual ignoring interposed non-stereogenic centers.

$$
\begin{array}{c}
\text{CH}_2\text{OH} \\
| \\
\text{C}=\text{O} \\
| \\
\text{HO}-\text{C}-\text{H} \\
| \\
\text{C}=\text{O} \\
| \\
\text{H}-\text{C}-\text{OH} \\
| \\
\text{CH}_2\text{OH}
\end{array} \Bigg\} \text{D–}\textit{threo}
$$

$$
\begin{array}{c}
\text{CH}_2\text{OH} \\
| \\
\text{H}-\text{C}-\text{OH} \\
| \\
\text{HO}-\text{C}-\text{H} \\
| \\
\text{C}=\text{O} \\
| \\
\text{C}=\text{O} \\
| \\
\text{HO}-\text{C}-\text{H} \\
| \\
\text{HO}-\text{C}-\text{H} \\
| \\
\text{CH}_2\text{OH}
\end{array} \Bigg\} \text{L–}\textit{altro}
$$

$$
\begin{array}{c}
\text{CH}_2\text{OH} \\
| \\
\text{H}-\text{C}-\text{OH} \\
| \\
\text{C}=\text{O} \\
| \\
\text{H}-\text{C}-\text{OH} \\
| \\
\text{HO}-\text{C}-\text{H} \\
| \\
\text{C}=\text{O} \\
| \\
\text{H}-\text{C}-\text{OH} \\
| \\
\text{H}-\text{C}-\text{OH} \\
| \\
\text{CH}_2\text{OH}
\end{array}
$$

D–*gulo*

D–*glycero*

D-*threo*-Hexo-2,4-
diulose

L-*altro*-Octo-
4,5-diulose
(not: L-*talo*)

D-*glycero*-D-*gulo*-
Nono-3,6-diulose

5.3
Ketoaldoses (Aldoketoses, Aldosuloses)

Logical extrapolation of the above procedures furnishes the systematic
...os-n-ulose names of **ketoaldoses**. However, **dehydro names** are frequently preferred in a biochemical context.

$$
\begin{array}{c}
\text{HC}=\text{O} \\
| \\
\text{HO}-\text{C}-\text{H} \\
| \\
\text{C}=\text{O} \\
| \\
\text{H}-\text{C}-\text{OH} \\
| \\
\text{H}-\text{C}-\text{OH} \\
| \\
\text{CH}_2\text{OH}
\end{array} \Bigg\} \text{D–}\textit{arabino}
$$

$$
\begin{array}{c}
\text{HC}=\text{O} \\
| \\
\text{HO}-\text{C}-\text{H} \\
| \\
\text{H}-\text{C}-\text{OH} \\
| \\
\text{C}=\text{O} \\
| \\
\text{H}-\text{C}-\text{OH} \\
| \\
\text{HO}-\text{C}-\text{H} \\
| \\
\text{H}-\text{C}-\text{OH} \\
| \\
\text{CH}_2\text{OH}
\end{array} \Bigg\} \text{L–}\textit{galacto}
$$

D–*glycero*

D-*arabino*-Hexos-3-ulose
(3-Dehydro-D-altrose)

D-*glycero*-L-*galacto*-Octos-4-ulose
(4-Dehydro-D-*erythro*-D-*altro*-octose)

For the cyclic hemiacetal forms of these sugars the position of the ring-size
designator depends upon which carbonyl group participates in ring formation.

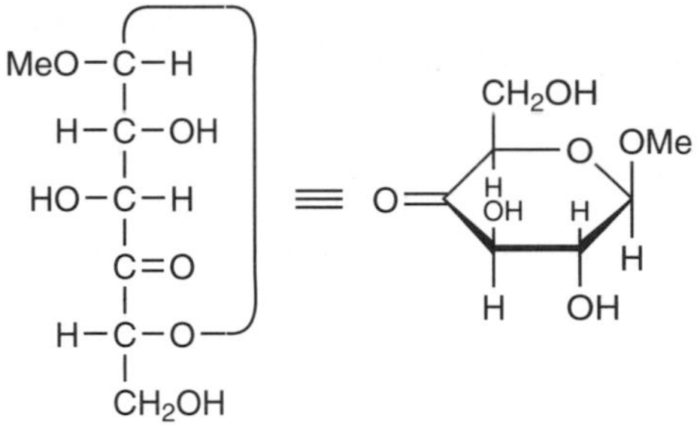

Methyl-β-D-*xylo*-Hexapy-
ranosid-4-ulose

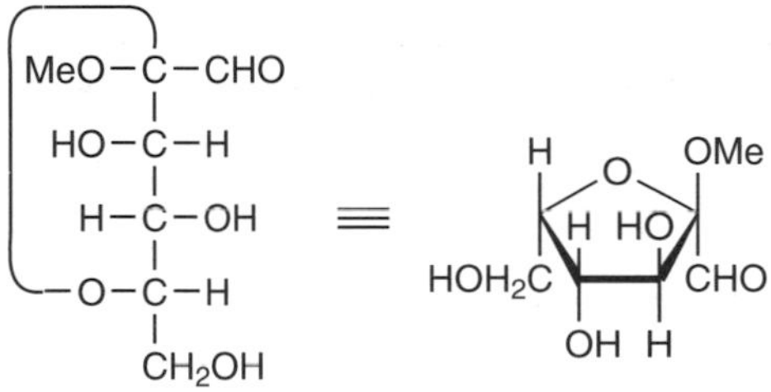

Methyl-α-L-*xylo*-Hexos-2-
ulo-2,5-furanoside

5.4
Deoxy Sugars

Deoxy prefixes with a frontal locant are considered as detachable, therefore ordered alphabetically and used indiscriminately both for trivially and for systematically named sugars. In assigning the pertinent configurational descriptors, interposed CH_2 (and also CO) groups are again disregarded.

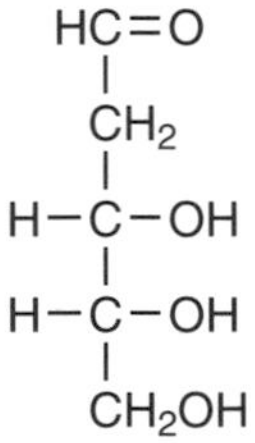

2-Deoxy-D-*ery-*
thro-Pentose
trad.: 2-Deoxyribose

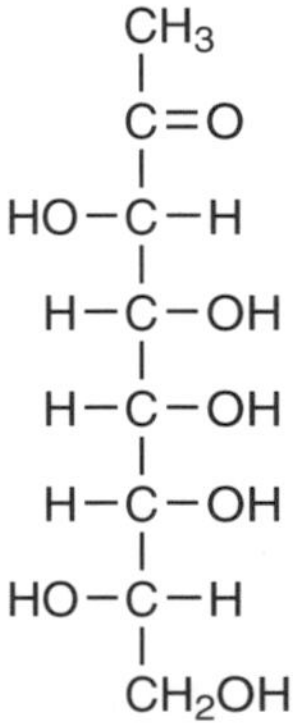

1-Deoxy-L-*glycero-*
D-*altro*-Oct-2-ulose

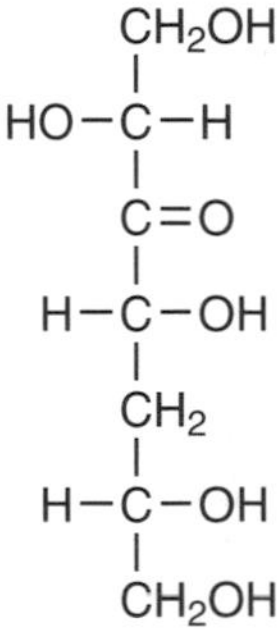

5-Deoxy-D-*arabino*-
Hept-3-ulose

5.5
Amino-Sugars and Analogously Substituted Derivatives

Amino-sugars are systematically named as amino derivatives of the corresponding deoxy-sugars where an amino group replaces a hydroxy group, also with respect to configuration. In the case of substitution at position 2 mostly the trivial ... osamine names are preferred, e.g.:

trivially: systematically:

D-Glucosamine   2-Amino-2-deoxy-D-glucose
D-Fucosamine   2-Amino-2,6-dideoxy-D-galactose
N-Acetyl-D-mannosamine   2-Acetamido-2-deoxy-D-mannose

Other non-terminal substituents are analogously accounted for, e.g.: 2-deoxy-2-C-phenyl-D-glucopyranose; 2,3-diazido-2,3-dideoxy-D-mannopyranose; 3-deoxy-3,3-dimethyl-D-*ribo*-hexose etc. Substitution of an aldehydic hydrogen is simply indicated by a C-substituent prefix: 1-C-phenyl-D-glucose.

5.6
Transformations of the Carbonyl Functions

5.6.1
Oximes, Hydrazones, Osazones

Sugar derivatives obtained by replacing carbonyl oxygen with nitrogen-linked groups retain their traditional names.

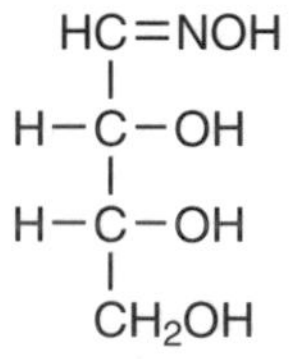

D-Erythrose oxime 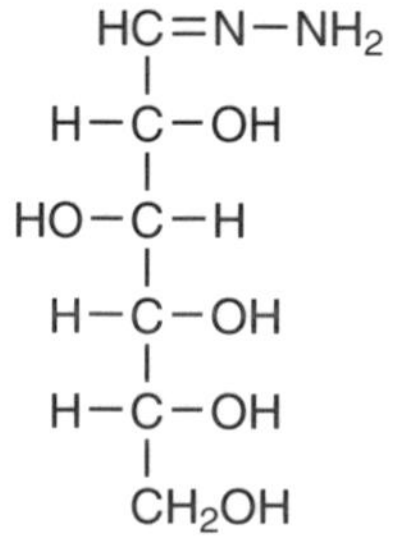   D-Glucose hydrazone 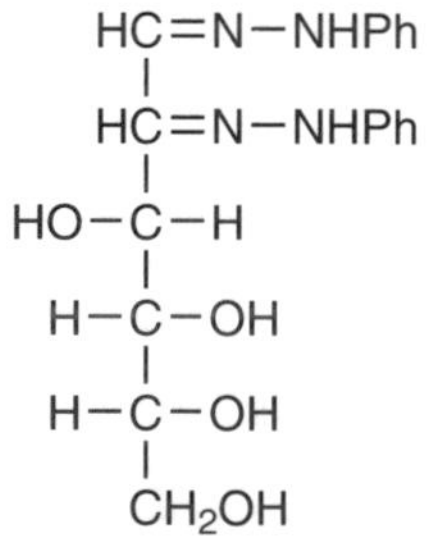   D-*arabino*-Hexos-2-ulose bis(phenylhydrazone) (D-Fructose phenylosazone)

5.6.2
Acetals, Ketals

Derivatives of sugars generated by transformation of the carbonyl group with alcohols are conventionally named as **(hemi)acetals, (hemi)ketals** and so on.

$$
\begin{array}{c}
\text{SEt} \\
|\\
\text{H}-\text{C}-\text{OH} \\
|\\
\text{H}-\text{C}-\text{OAc} \\
|\\
\text{AcO}-\text{C}-\text{H} \\
|\\
\text{H}-\text{C}-\text{OAc} \\
|\\
\text{H}-\text{C}-\text{OAc} \\
|\\
\text{CH}_2\text{OAc}
\end{array}
$$

(1S)-2,3,4,5,6-Penta-O-ace-tyl-D-glucose S-ethyl mo-nothiohemiacetal

$$
\begin{array}{c}
\text{CH}_2\text{OH} \\
|\\
\text{C(OEt)}_2 \\
|\\
\text{HO}-\text{C}-\text{H} \\
|\\
\text{H}-\text{C}-\text{OH} \\
|\\
\text{H}-\text{C}-\text{OH} \\
|\\
\text{CH}_2\text{OH}
\end{array}
$$

D-Fructose-diethyl ketal

$$
\begin{array}{c}
\text{HC} \underset{\text{S}}{\overset{\text{S}}{\big\langle}} \\
|\\
\text{H}-\text{C}-\text{OH} \\
|\\
\text{H}-\text{C}-\text{OH} \\
|\\
\text{H}-\text{C}-\text{OH} \\
|\\
\text{H}-\text{C}-\text{OH} \\
|\\
\text{CH}_2\text{OH}
\end{array}
$$

D-Allose-propane 1,3-diyl dithioacetal

5.7
Branched Sugars

Branched monosaccharides are treated according to the afore-stated principles as substituted unbranched sugars, with customary trivial names being retained. The chief criteria for the selection of parent saccharides are:

a) the highest ranked functional type: aldaric acid > uronic acid/keto-aldonic acid/aldonic acid > dialdose > ketoaldose/aldose > diketose > ketose,

b) the longest chain: heptose > hexose,

c) the alphabetically preferred parent system or configurational prefix: allose > altrose; *gluco* > *gulo*,

d) the preferred configurational symbol: D > L; $\alpha > \beta$,

e) the parent system containing the largest number of substituents expressed as prefixes.

```
        HC=O
         |                         HC=O                            HC=O
HOH2C—C—OH                          |                               |
         |                  H—C—OH                          H—C—OH ⎤
   H—C—OH                           |                               |  ⎬ D–erythro
         |           HOH2C—C—OH                            H—C—OH ⎦
   H—C—OH                           |                               |
         |                     CH2OH                     HOH2C—C—OH
      CH2OH                                                        |
                                                               CH2OH
```

2-*C*-(Hydroxyme-   3-*C*-(Hydroxyme-   4-*C*-(Hydroxymethyl)-
thyl)-D-ribose   thyl)-D-*glycero*-   D-*erythro*-pentose
(Hamamelose)   tetrose   (The achiral C-4
                         (D-Apiose)   is ignored)

```
                HC=O
                 |
          H—C—OH
                 |
          H—C—OH    OH
                 |   ╱
          H—C—C···H
                 |    ╲OH
   L–talo  HO—C—H  C···
                 |   | ''H
              C=O  CH2OH
                 |
              CH2OH
```

```
                        HC=O
                         |              CH2OH
                  H—C—OH           |
                         |         H—C—CH2—CH—CH2OH
                  H—C—CH2
                         |
           H3C—C—H
                 |
           HO—C—H
                 |
              CH2OH
```

4-Deoxy-4-[(1*R*,2*S*) or (L-*ery-*   3,4-Dideoxy-3-[3-hydroxy-2-
thro)-1,2,3-trihydroxypropyl]-L-   (hydroxymethyl)propyl]-4-*C*-
talo-heptos-6-ulose   methyl-L-mannose

5.8
Sugar Alcohols (Alditols)

Alkanepolyols obtained by reduction of sugars are generally characterized
by the systematic ending ...**itol**, e.g.: erythritol, ribitol, mannitol, etc.
Besides, these names are generally derived from the most senior (see preceding paragraph) parent sugar, e.g.: D-arabinitol (not D-lyxitol), D-glu-
citol (not L-gulitol); the traditional name sorbitol is no longer recommended.

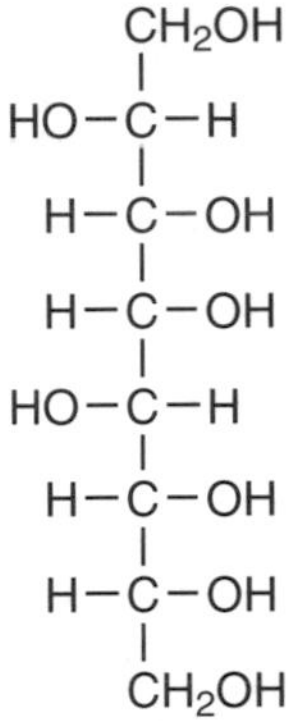

$$
\begin{array}{c}
CH_2OH \\
| \\
HO-C-H \\
| \\
H-C-OH \\
| \\
H-C-OH \\
| \\
HO-C-H \\
| \\
H-C-OH \\
| \\
H-C-OH \\
| \\
CH_2OH
\end{array}
$$

D-*erythro*-L-*ga-lacto*-Octitol

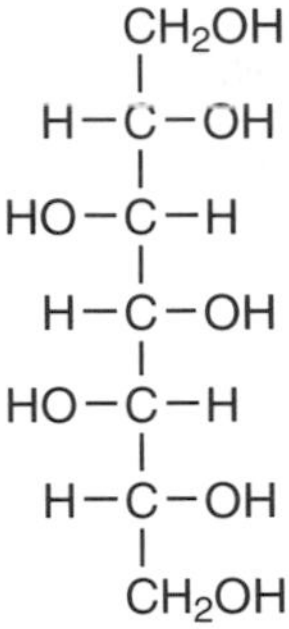

$$
\begin{array}{c}
CH_2OH \\
| \\
H-C-OH \\
| \\
HO-C-H \\
| \\
H-C-OH \\
| \\
HO-C-H \\
| \\
H-C-OH \\
| \\
CH_2OH
\end{array}
$$

meso-D-*glycero*-L-*ido*-Heptitol

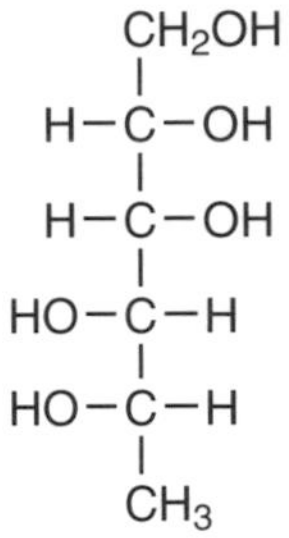

$$
\begin{array}{c}
CH_2OH \\
| \\
H-C-OH \\
| \\
H-C-OH \\
| \\
HO-C-H \\
| \\
HO-C-H \\
| \\
CH_3
\end{array}
$$

L-Rhamnitol or
1-Deoxy-L-mannitol

5.9
Acids Derived from Sugars

Through oxidation of the terminal aldehyde and alcohol functions of monosaccharides four different types of polyhydroxycarboxylic acids can be generated for which numerous traditional names are in use (see Table 17). Systematic names for these compounds, though, can very simply be formed by appropriate transformations or replacements of the ... ose and ... ulose endings as shown below:

a) oxidation of the aldehyde function: ... **onic acid** (generic name: **aldonic acids**),
b) oxidation of the terminal alcohol function: ... **uronic acid** (generic name: **uronic acids**),
c) oxidation of both terminal functions: ... **aric acid** (generic name: **aldaric acids**),
d) oxidation of *one* terminus of ketoses: ... **ulosonic acid** (generic name: **ketoaldonic acids**).

Derivatives of these acids are named in accordance with the general rules of substitutive nomenclature.

a)

```
      CO₂H                    CO₂Me                  O=C————
       |                        |                      |      |
  H — C — OH              H — C — NH₂             H — C — OH  |
       |                        |                      |      |
 HO — C — H               H — C — OH                  CH₂     |
       |                        |                      |      |
  H — C — OH              HO — C — H              H — C — OH  |
       |                        |                      |      |
  H — C — OH              HO — C — H              H — C — O———
       |                        |                      |
     CH₂OH                    CH₂OH                   CH₂OH
```

D-Gluconic acid   Methyl 2-amino-2-   3-Desoxy-D-*ribo*-hex-
                         deoxy-L-mannonate   ono-1,5-lactone

b)

```
      HC=O                 ┌— H — C — OH            HC=O
       |                   |        |               |
  H — C — OH               |   HO — C — H      H — C — OH
       |                   |        |               |
 HO — C — H               |   HO — C — H      H — C — OH
       |                   |        |               |
 HO — C — H               |    H — C — O——     H — C — O———
       |                   |        |               |       |
  H — C — OH               |    H — C — OH     H — C — OH   |
       |                   |        |               |       |
     CO₂H                 └——     CO₂Et       H — C — OH   |
                                                    |       |
                                                   O=C——————
```

D-Galacturonic acid   Ethyl α-D-*manno*-   D-*glycero*-D-*allo*-Hep-
                         furanuronate   turono-4,6-lactone

c)

```
                               CONH₂                ┌——— C=O
      CO₂H                       |                  |     |
       |                    H — C — OH              |  H — C — OH
  H — C — OH                     |                  |     |
       |                    HO — C — H              |  H — C — O—┐
  H — C — OH                     |                  |     |      |
       |                    H — C — OH             └— O — C — H  |
     CO₂H                        |                        |      |
                            H — C — OH               HO — C — H  |
                                 |                        |      |
                               CO₂H                      O=C—————
```

meso- or 2R,3S-   D-Glucar-1-amic   L-Mannaro-1,4:
Erythraric acid   acid   6,3-dilactone
trad.: *meso*-Tartaric
acid

d)

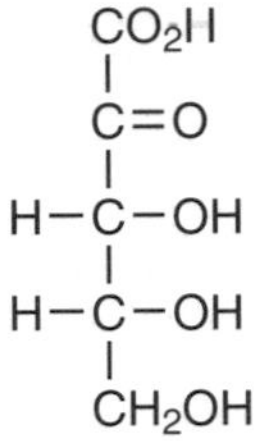

CO₂H
|
C=O
|
H—C—OH
|
H—C—OH
|
CH₂OH

D-*erythro*-Pent-
2-ulosonic acid

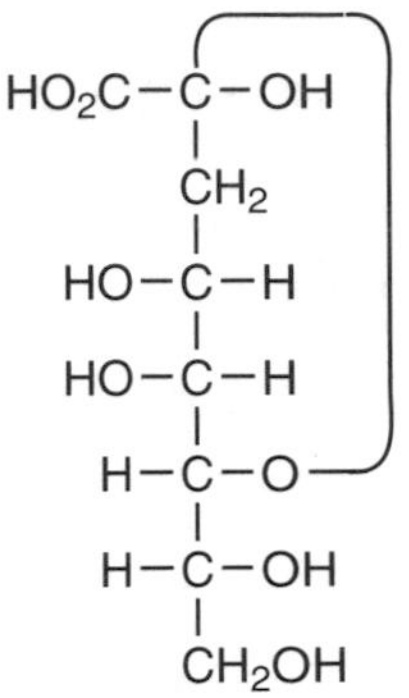

HO₂C—C—OH
|
CH₂
|
HO—C—H
|
HO—C—H
|
H—C—O
|
H—C—OH
|
CH₂OH

3-Deoxy-α-D-*manno*-
oct-2-ulosonic acid

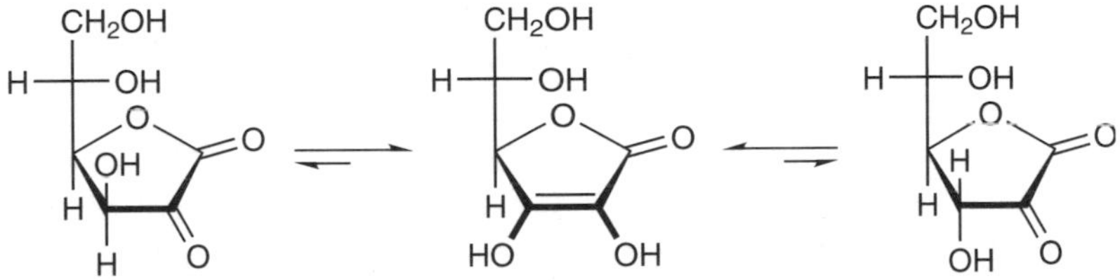

L-*xylo*-Hex-2-ulo-
sono-1,4-lactone

L-*threo*-Hex-2-en-
ono-1,4-lactone

L-*lyxo*-Hex-2-ulo-
sono-1,4-lactone

L-Ascorbic acid

5.10
O-Substitution

5.10.1
O-Substitution with Alkyl and Acyl Groups

O-Alkyl and *O*-acyl derivatives of saccharides can most easily be named as such even though traditional ester names are still widely used for the latter.

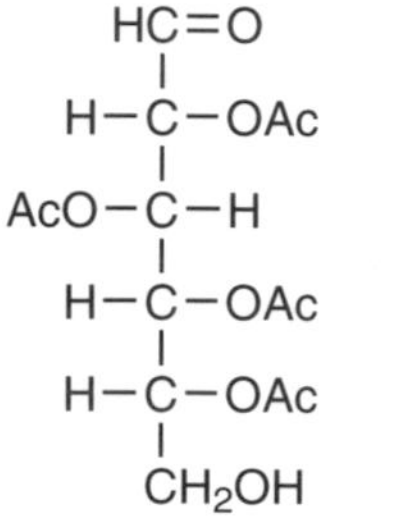

Penta-*O*-acetyl-*aldehydo*-D-glu-cose or: *Aldehydo*-D-glu-cose pentaacetate

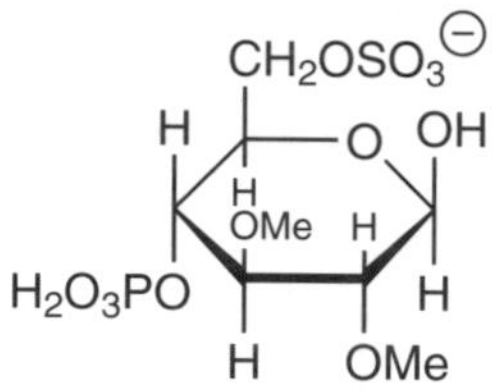

2,3-Di-*O*-methyl-4-*O*-phosphono-6-*O*-sulfonato-β-D-gluco-pyranose

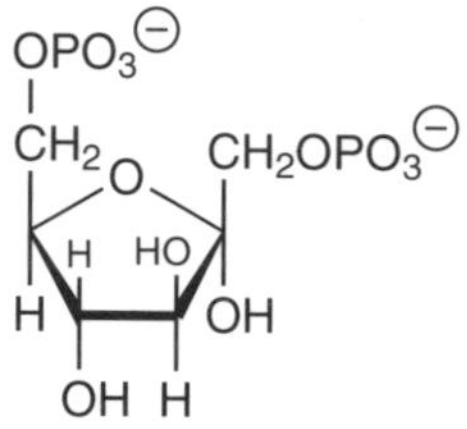

1,6-Di-*O*-phosphonato-α-D-fructofuranose or: 1,6-Bisphospho-α-D-fructofuranose or: α-D-Fructofu-ranose 1,6-bisphosphate

The inevitable acronym **ADP** too comes from the ester name adenosine-5'-diphosphate customarily preferred to the systematic denominations 5'-*O*-(diphosphonato)adenosine or 5'-(diphospho)adenosine.

5.10.2
Cyclic Acetals and Ketals

Cyclic acetals and ketals produced from saccharides by reactions with aldehydes and ketones, respectively, are identified as **x,y-*O*-alkyl(id)ene** derivatives.

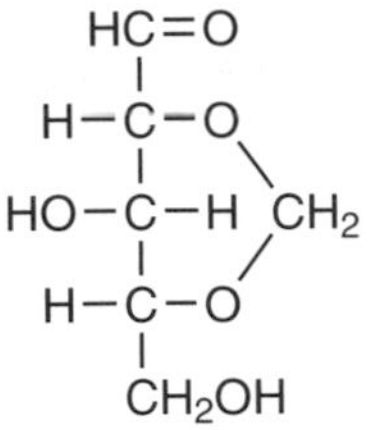

2,4-*O*-Methyl-
enexylose

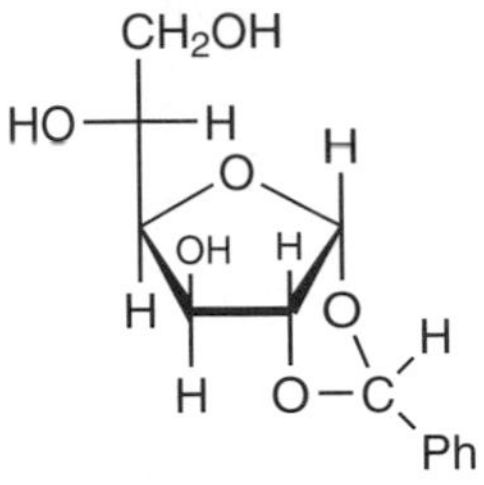

1,2-*O*-Benzylidene-
D-glucofuranose

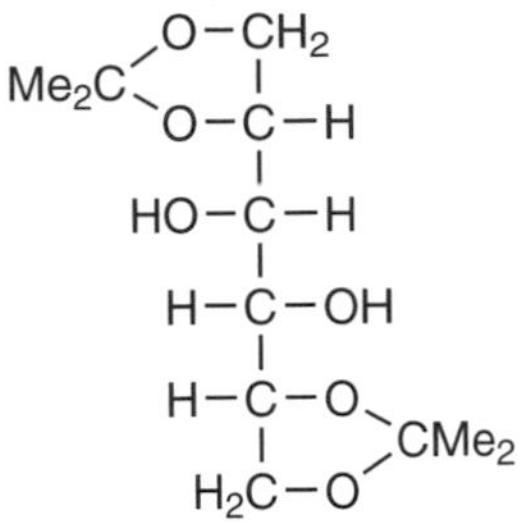

1,2:5,6-Di-*O*-isoprop-
ylidene-D-mannitol

5.11
Monosaccharides as Substituent Groups

If need arises, five types of substituent groups can be derived from sugars and their derivatives, whose names are formed by supplementing the group suffix ... osyl with suitable descriptors.

a) the free valence is generated through detachment of the anomeric hydroxy group: ...osyl. These are the prototypical glycosyl groups found in the general literature (see also the next section).
b) the free valence at C-1 is generated while keeping the hydroxy group in place: **1-hydroxy ... osyl.**
c) a free valence is generated at any (non-anomeric) C atom bearing a hydroxy group: **...os-n-C-yl.**
d) a free valence is generated at any (non-anomeric) hydroxy group: **...os-n-O-yl.**
e) the free valence is generated at a reduced center: **n-deoxy ... os-n-yl.**

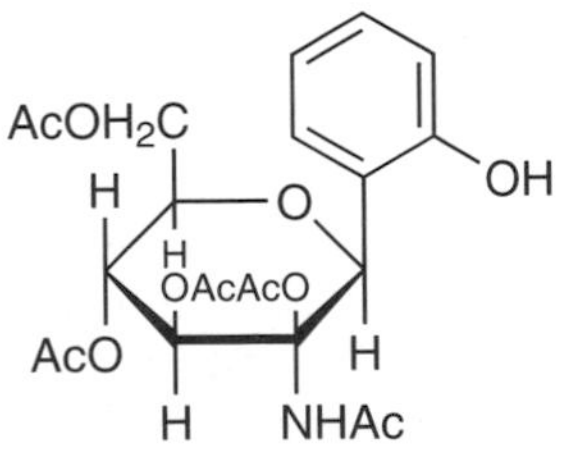

a) 2-(2-*C*-Acetamido-2,3,4,6-
tetra-*O*-acetyl-β-D-manno
pyranosyl)phenol

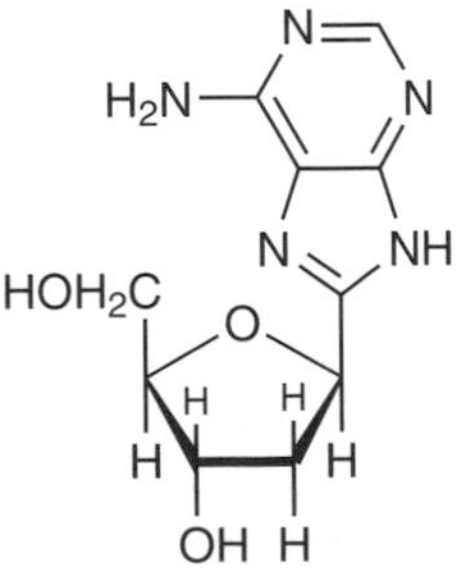

a) 8-(2-Deoxy-β-D-*erythro*-
pentofuranosyl)adenine

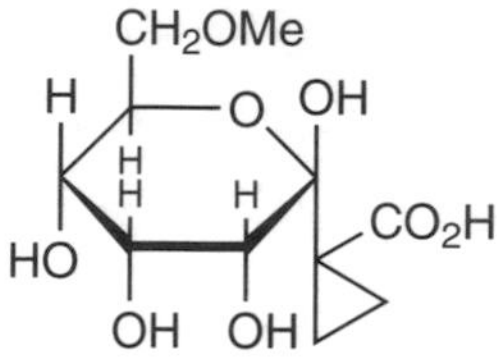

b) 1-(1-Hydroxy-6-O-methyl-
α-D-allopyranosyl)cyclo-
propane-1-carboxylic acid

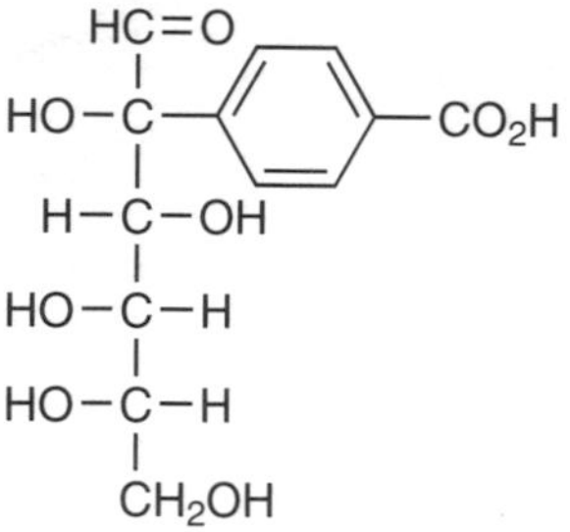

c) 4-(L-Glucos-2-C-yl)-benzoic acid

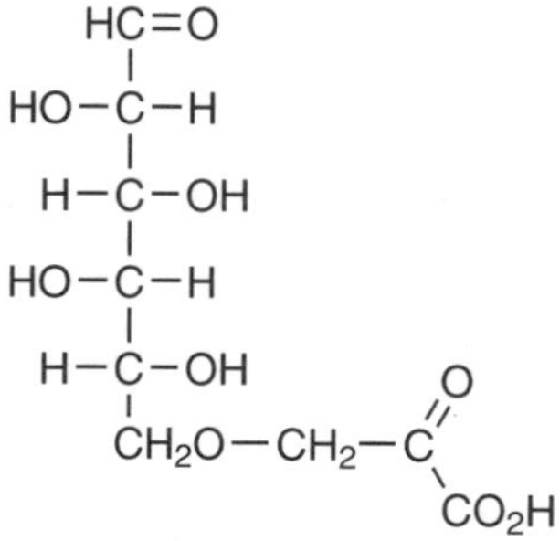

d) (D-Idos-6-O-yl)pyruvic
acid or:
6-O-(Oxalomethyl)-D-idose

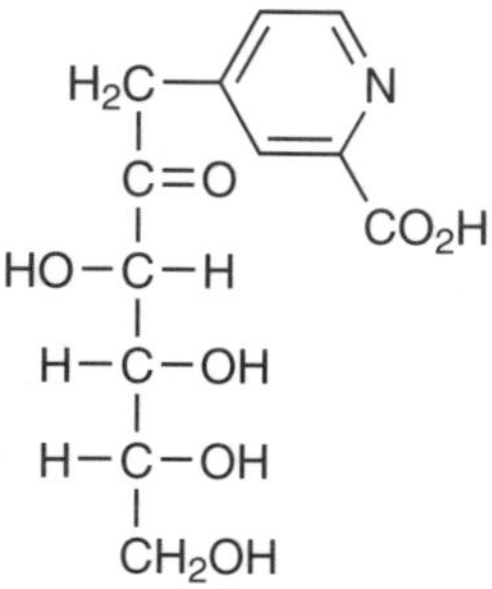

e) 4-(1-Desoxy-D-fructos-1-yl)-
pyridine-2-carboxylic acid or:
1-(2-Carboxypyridin-4-yl)-1-
deoxy-D-fructose

5.12
Glycosides and Glycosyl Compounds

5.12.1
Glycosides

Mixed acetals and ketals where the anomeric hydroxy group of the cyclic forms of monosaccharides is substituted by alkoxy, alkylthio/seleno groups are covered by the generic terms **glycoside, thio/selenoglycoside**. Three naming variants again exist here, about whose relative weight no more can be said than that the traditional ...**oside** names are generally preferred at least for simple cases.

a) transformation of the ...**ose** name into ...**yl**...**oside** or ...**yl**...-**n**-thio...**oside**.

b) employment of ... **osyloxy** and ... **osylthio prefixes** if the saccharide functions as substituent group.

c) use of the format **O/S/Se...osyl** for the glycosyl group substituting a certain hydroxy/thio/seleno compound.

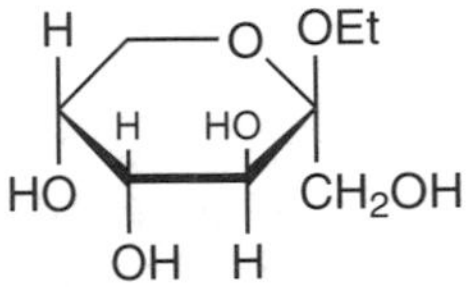

a) Ethyl β-D-fructo-
pyranoside

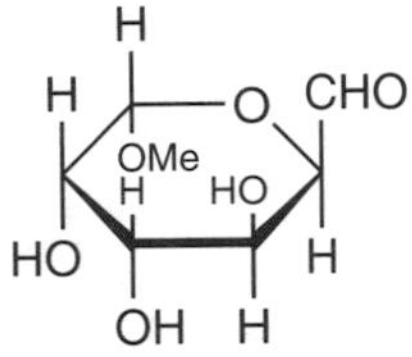

a) Methyl (6R)-D-
gluco-hexodialdo-
6,2-pyranoside

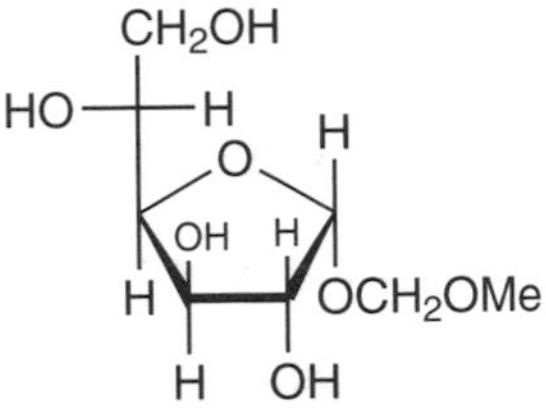

a) (Methoxymethyl)
α-D-glucofuranoside

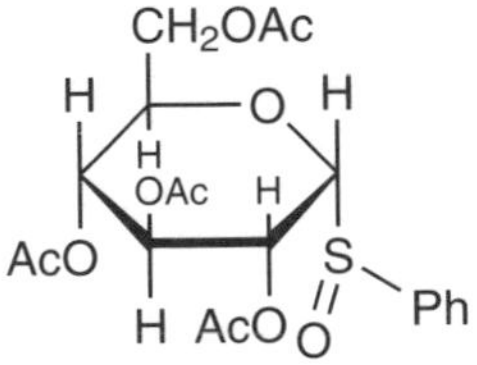

a) Phenyl tetra-O-acet-
yl-1-thio-α-D-gluco-
pyranoside S-oxide
or: Phenyl tetra-
O-acetyl-α-D-gluco-
pyranosyl sulfoxide

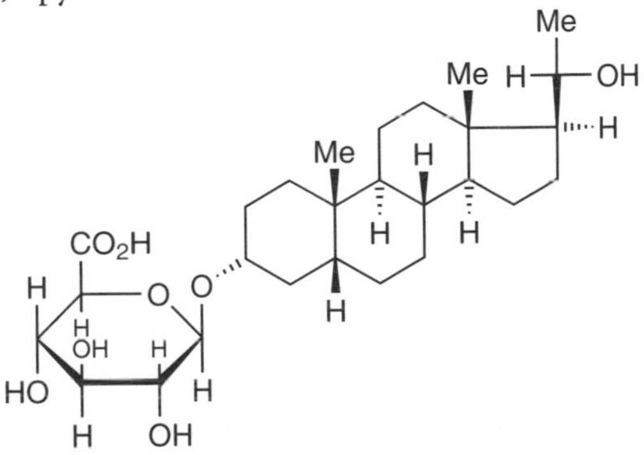

a) (20S)-20-Hydroxy-5-β-pregnan-3α-
yl β-D-glucopyranosiduronic acid
b) (20S)-3α-(β-D-Glucopyranosyloxy-
uronic acid)-5β-pregnan-20-ol
(for biochemical usage:
pregnanediol 3-glucuronide)

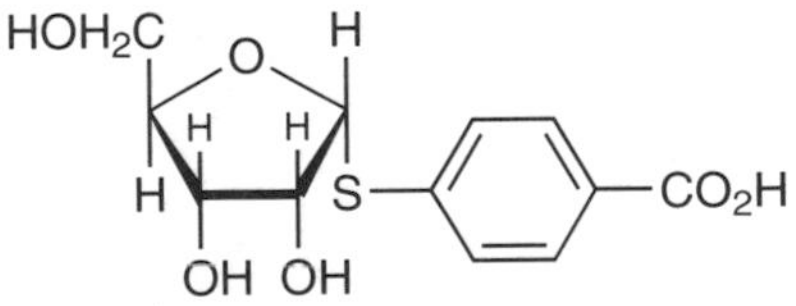

a) 4-Carboxyphenyl 1-thio-α
-D-ribofuranoside
b) 4-(α-D-Ribofuranosylthio)
benzoic acid

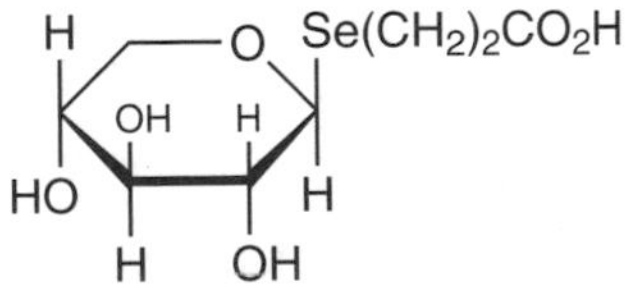

a) 2-Carboxyethyl 1-seleno-
β-D-xylopyranoside
b) 3-(β-D-Xylopyranosylsele-
no)propanoic acid

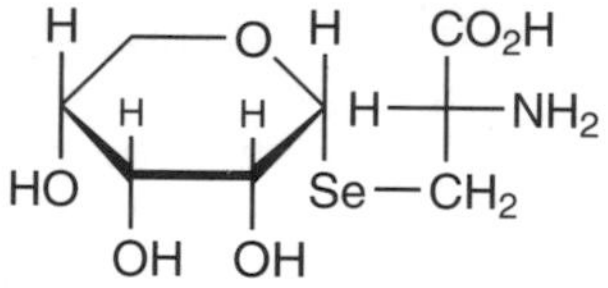

a) (*S*)-2-Amino-2-carboxyethyl 1-seleno-
   α-D-ribopyranoside
b) 3-(α-D-Ribopyranosylseleno)-D-alanine
c) *Se*-α-D-Ribopyranosyl-D-selenocysteine

5.12.2
Glycosyl Compounds

Compounds in which the anomeric hydroxy group has been substituted by
halogen atoms, pseudohalogen or amino groups are named **quasi radicofunctionally** as glycosyl derivatives of the respective functional classes.

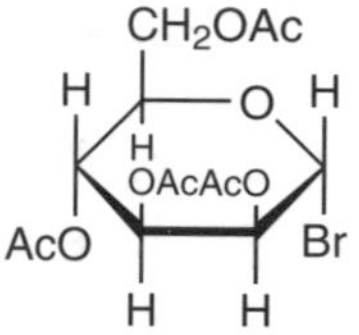

Tetra-*O*-acetyl-α-D-
mannosyl bromide

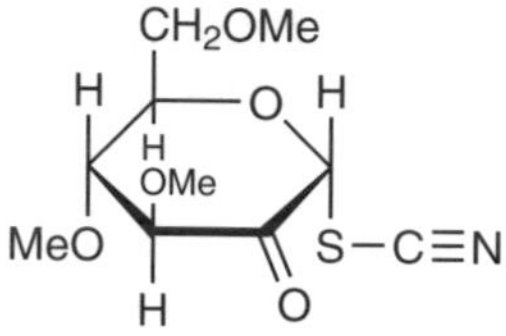

3,4,6-Tri-*O*-methyl-α-D-*arabino*-hexo-
pyranosyl-2-ulose thiocyanate

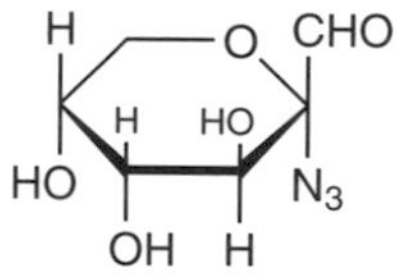

α-D-*arabino*-Hexos-2-
ulo-2,6-pyranosyl azide
or:
aldehydo-α-D-*arabino*-He-
xos-2-ulopyranosyl azide

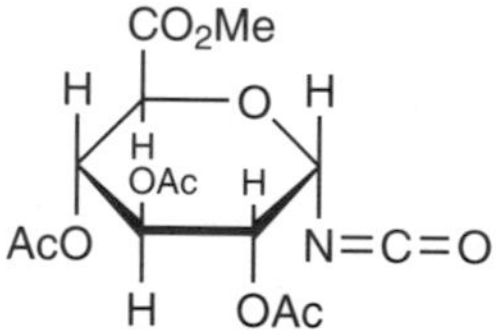

Methyl-(2,3-4-tri-*O*-acetyl-α-D-gluco-
pyranosyl)uronate isocyanate
systematically more logical would be:
Methyl-(2,3,4-tri-*O*-acetyl-1-isocyanato-
1-deoxy-α-D-glucopyranuronate

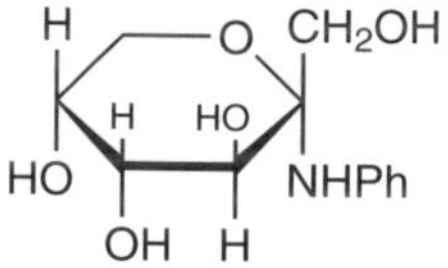

N-Phenyl-α-D-fructo-
pyranosylamine

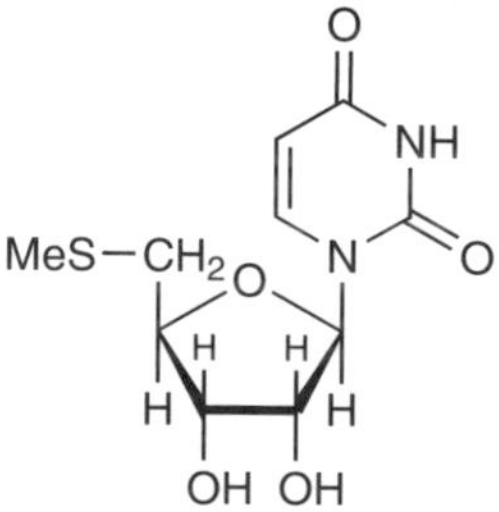

1-(5-*S*-Methyl-5-thio-β-D-ribofuranosyl)
uracil

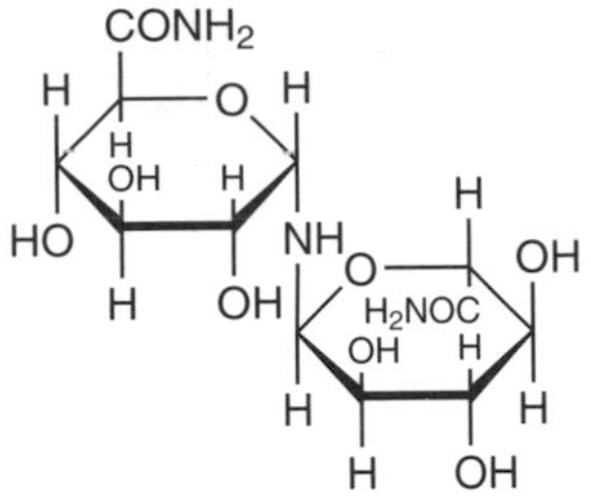

Bis(α-D-glucopyranosyluronamid)amine
or:
1,1′-Iminobis-(1-deoxy-α-D-glucopyra-
nuronamide)

C-Glycosyl compounds have already been dealt with in Section 5.11.

5.13
Oligosaccharides

5.13.1
Oligosaccharides with Free Hemiacetal Group

These compounds are generally named **glycosyl[glycosyl]$_n$ glycoses** with
the connectivity locants placed in round brackets between the individual
component terms. The same format is used for the specific short formulae
customary in carbohydrate chemistry. For disaccharides of that type Chem.
Abstr. still uses the traditional *O*-locants in front of the full name.

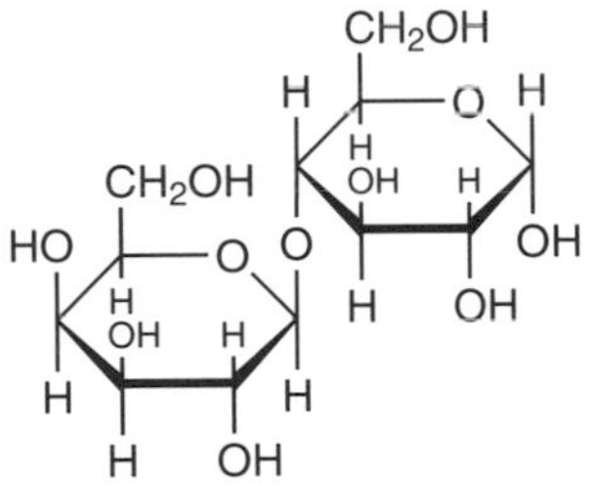

β-D-Galactopyranosyl-(1 → 4)-α-D-
glucopyranose
[β-D-Gal*p*-(1 → 4)-α-D-Glc*p*]

Chem. Abstr.:

4-*O*-β-D-Galactopyranosyl-α-D-gluco-
pyranose
triv.: α-Lactose

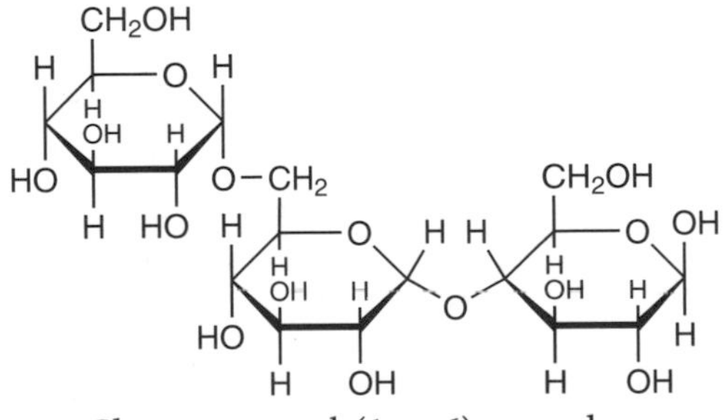

α-D-Glucopyranosyl-(1 → 6)-α-D-glucopyranosyl-(1 → 4)-β-D-gluco-
pyranose [α-D-Glc*p*-(1 → 6)-α-D-Glc*p*-(1→ 4)-β-D-Glc*p*]

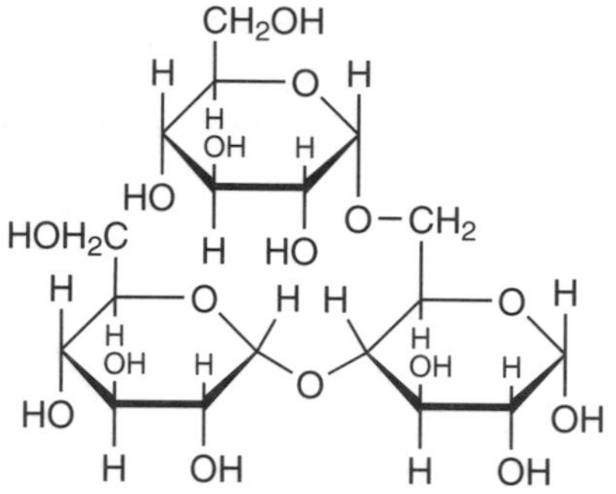

α-D-Glucopyranosyl-$(1 \rightarrow 4)$-[α-D-glucopyranosyl-$(1 \rightarrow 6)$]α-D-glucopyranose [α-D-Glc$p$-$(1 \rightarrow 4)$[α-D-Glc$p$-$(1 \rightarrow 6)$]-α-D-Glc$p$] or: 4,6-Di-O-(α-D-glucopyranosyl)-α-D-glucopyranose

5.13.2
Oligosascharides without Free Hemiacetal Group

Oligosaccharides linked exclusively through their anomeric hydroxy groups can generally be named in a **sequential manner** as **glycosyl(glycosyl)$_n$ glycosides**. For trisaccharides in particular a second naming method is in use, **centering** on the **most senior individual component** potentially located in the interior.

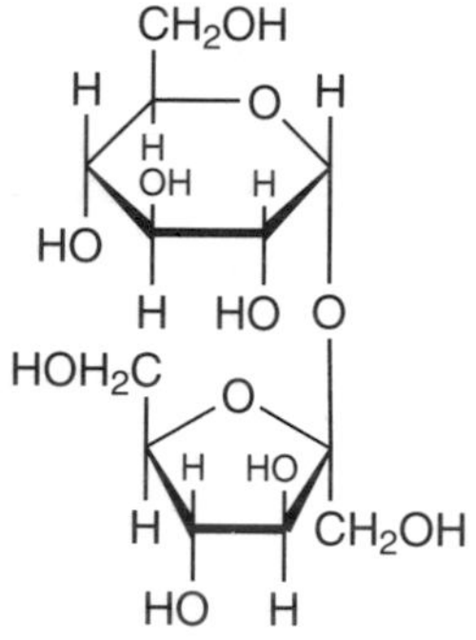

β-D-Fructofuranosyl α-D-glucopyranoside [β-D-Fru$f$-$(2 \leftrightarrow 1)$-α-D-Glc$p$] (trivially: Sucrose, Saccharose)

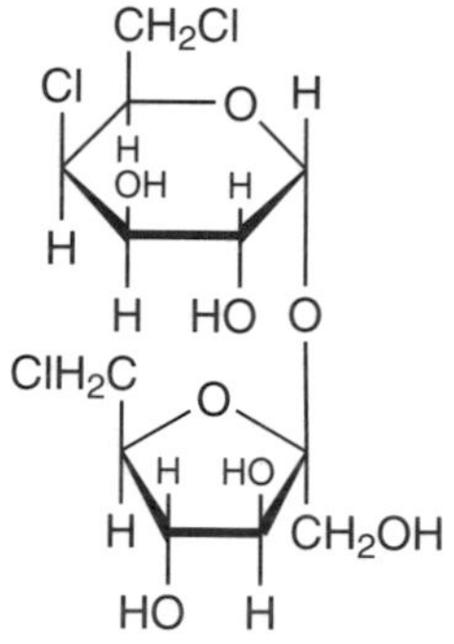

6-Chloro-6-deoxy-β-D-fructofuranosyl 4,6-dichloro-4,6-dideoxy-α-D-galactopyranoside

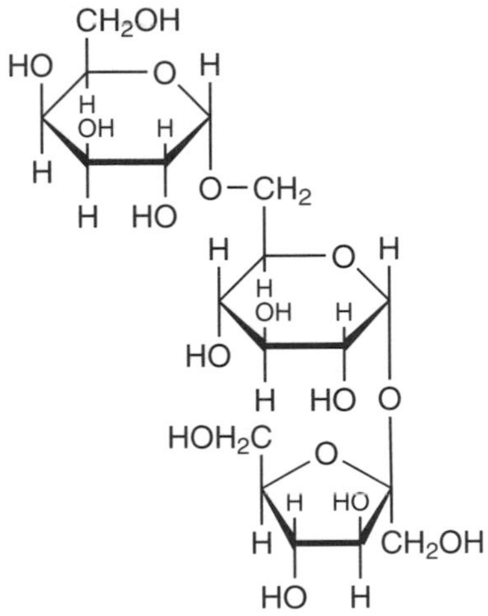

sequentially:
α-D-Galactopyranosyl-(1 → 6)-
α-D-glucopyranosyl β-D-fructo-
furanoside
[α-D-Gal*p*-(1 → 6)-α-D-Glc*p*-
(1 ↔ 2)-β-D-Fru*f*]
related to Glucopyranose:
β-D-Fructofuranosyl α-D-galacto-
pyranosyl-(1 → 6)-α-D-glucopyra-
noside (trivially: Raffinose)

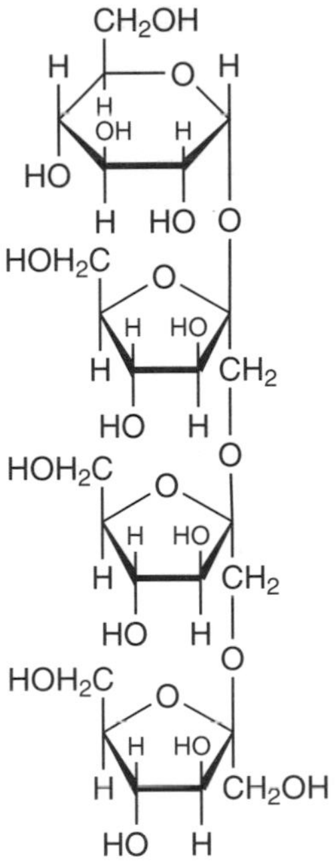

β-D-Fructofuranosyl-(2 → 1)-
β-D-fructofuranosyl-(2 → 1)-
β-D-fructofuranosyl-α-D-gluco-
pyranoside
(trivially: Nystose)
[β-D-Fru*f*-(2 → 1)-β-D-Fru*f*-
(2 → 1)-β-D-Fru*f*-(2 ↔ 1)-α-D-
Glc*p*]

5.13.3
Polysaccharides (Glycans)

Polysaccharides, finally, are subsumed under the generic term **glycan**
whereas structurally defined representatives of this compound class are
characterized by the ending ... **an** which should also be used when creat-
ing new trivial names in this field. For **homopolysaccharides** the ... **ose**
ending is simply changed to ... **an**; for homopolysaccharides **substituted in
a defined manner** it is the ... ose ending of the backbone-glycose which is
to be changed to ... an, if needed with a connectivity descriptor preceding
the glycan name: **xylose → xylan, mannose → mannan,** etc.

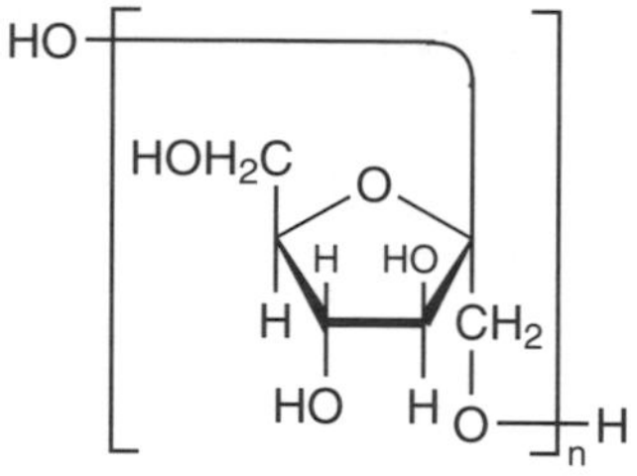

$(2 \rightarrow 1)$-β-D-Fructofuranan

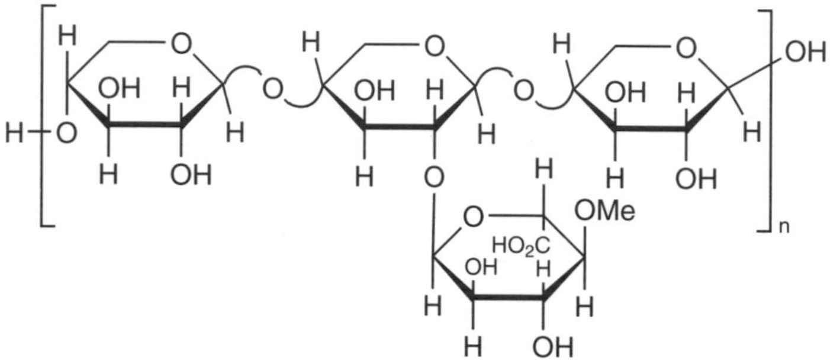

$(4$-O-Methyl-α-D-glucurono$)$-D-xylan

5.14
Customary Trivial Names

To make the chapter on carbohydrate nomenclature complete, Table 17 supplements a compilation of a larger number of **widely used trivial names** of derivatized saccharides with their systematic equivalents.

Table 17. Frequently used trivial names for derivatives of saccharides

Abequose (Abe)	3,6-Didesoxy-D-*xylo*-hexose
Amylose	$(1 \rightarrow 4)$-α-D-Glucopyranan
Apiose (Api)	3-*C*-(Hydroxymethyl)-*glycero*-tetrose
Ascorbic acid	L-*threo*-Hex-2-enono-1,4-lactone
Cellobiose, etc.	β-D-Glucopyranosyl-$(1 \rightarrow 4)$-D-glucose, etc.
Cladinose	2,6-Didesoxy-3-*C*-methyl-3-*O*-methyl-L-*ribo*-hexose
2-Deoxyribose (dRib)	2-Deoxy-*erythro*-pentose
2-Deoxyglucose (2dGlc)	2-Deoxy-*arabino*-hexose
Digitalose	6-Deoxy-3-*O*-methyl-D-galactose
Digitoxose	2,6-Dideoxy-D-*ribo*-hexose

Table 17 (continued)

Fucosamine (FucN)	2-Amino-2,6-dideoxygalactose
Fucose (Fuc)	6-Deoxygalactose
Galactosamine (GalN)	2-Amino-2-deoxygalactose
Gentiobiose	β-D-Glucopyranosyl-(1 → 6)-D-glucose
Glucosamine (GlcN)	2-Amino-2-deoxyglucose
Glucosaminitol (GlcN-ol)	2-Amino-2-deoxyglucitol
Glyceraldehyde	2,3-Dihydroxypropanal
Glycerol (Gro)	Propane-1,2,3-triol
Glycerone (1,3-Dihydroxyacetone)	1,3-Dihydroxypropanone
Hamamelose	2-C-(Hydroxymethyl)-D-ribose
Inulin	(2 → 1)-β-D-Fructofuranan
Lactose (Lac)	β-D-Galactopyranosyl-(1 → 4)-D-glucose
Maltose	α-D-Glucopyranosyl-(1 → 4)-D-glucose
Mannosamine (ManN)	2-Amino-2-deoxymannose
Melibiose	α-D-Galactopyranosyl-(1 → 6)-D-glucose
Neuraminic acid	5-Amino-3,5-dideoxy-D-*glycero*-D-*galacto*-non-2-ulosonic acid
Raffinose	β-D-Fructofuranosyl-(α-D-galactopyranosyl-(1 → 6)-α-D-glucopyranoside)
Rhamnose (Rha)	6-Deoxymannose
Sucrose, Saccharose	β-D-Fructofuranosyl-α-D-glucopyranoside
Mucic acid	*meso*-Galactaric acid
Sedoheptulose	D-*altro*-Hept-2-ulose
Streptobiose	2-Deoxy-2-methylamino-α-L-glucopyranosyl-(1 → 2)-5-deoxy-3-C-formyl-L-lyxose
Streptose	5-Deoxy-3-C-formyl-L-lyxose
Trehalosamine	2-Amino-2-deoxy-α-D-glucopyranosyl-α-D-glucopyranoside
α,α-Trehalose	α-D-Glucopyranosyl-α-D-glucopyranoside
Tartaric acid	Erythraric/Threaric acid
Saccharic acid	D-Glucaric acid

6 Construction of the Names of Complex Compounds

Chapters 1, 2, and 3 of this book have dealt in some detail with the nomenclature rules for parent structures and the wealth of different functional compound classes based there upon. The present chapter will recapitulate once again in a more summary manner the most important directives for the construction of the complete names of more complex compounds. This entails a reconsideration of the question of priorities for ring and chain systems already briefly approached in the pertinent sections of Chapter 1. Since functional groups must ultimately be included in the decisions, extension of the priority rules becomes imperative, as will be shown in the following sections.

6.1
Determination of the Highest Ranked Chain (Main or Senior Chain)

For acyclic compounds the definition of the highest ranked (main, senior) chain concomitantly determines that chain on which the name is to be based. For its selection the following criteria are applied sequentially: if the preceding criterion is inconclusive the next one comes in force.

a) greatest number of most senior characteristic groups,
b) largest number of heteroatoms (for hetero chains named with "a" nomenclature),
c) largest number of multiple bonds taken together,
d) greatest length,
e) for hetero chains: largest number of most senior hetero atoms (Tables 4, 20),
f) largest number of double bonds,
g) for hetero chains: lowest locants for all heteroatoms taken together, then according to their seniorities (Tables 4, 20),
h) lowest locants for the most senior characteristic groups expressed as suffixes,
i) lowest locants for all multiple bonds taken together,
j) lowest locants for double bonds,

k) largest number of substituents expressed as prefixes,
l) lowest locants for all prefix substituents of the main chain,
m) most senior prefix substituent according to alphabetical order,
n) lowest locants for the most senior prefix substituents.

6.2
Determination of the Most Senior Ring System

The following criteria are examined sequentially until a decision is reached.

a) largest number of most senior characteristic groups expressed as suffixes,
b) for heterocycles the ranking principles of Section 1.2.2.3 apply:
   ba) heterocycles are preferred to carbocycles,
   bb) nitrogen-containing rings,
   bc) most senior hetero element according to the "a"-term Tables 4, 20,
   bd) the ring system with the largest number of rings,
   be) the ring system having the largest ring,
   bf) the ring having the largest number of hetero elements,
   bg) the ring displaying the greatest variety of hetero elements,
   bh) the ring containing the largest number of senior elements,
   bi) the ring with the lowest possible set of locants before fusion,
c) for carbocycles too, the system containing the largest number of rings is to be preferred,
d) the system with the largest individual ring when the first point of difference is evaluated,

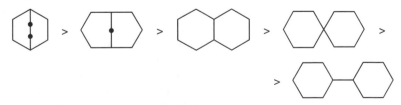

e) the system containing the largest number of atoms common to two or more rings,

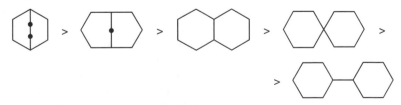

f) lowest letters for fusion positions: **naphtho[2,1-*f*]quinoline** > **naphtho[1,2-*g*]quinoline**,

g) lowest numerals for fusion or linking positions, e. g.:
    naphtho[1,2-*f*]quinoline > naptho[2,1-*f*]quinoline > ... [2,3-*f*] ...
    tricyclo[5.3.1.0^{2,4}]undecane > tricyclo[5.3.1.0^{3,5}]undecane
    spiro[cyclopentane-1,1'-indene] > spiro[cyclopentane-1,2'-indene]
    2,3'-bipyridine > 3,3'-bipyridine
h) lowest degree of hydrogenation

i) lowest locants for indicated hydrogen,
j) lowest locants for free valences,
k) lowest locants for characteristic groups expressed as suffixes,
l) largest number of substituents expressed as prefixes,
m) lowest locants when all prefix substituents, hydro prefixes, "ene"and
    "yne"positions are considered together,

3-Chloro-1,2-dihydro-   2-Chloro-2,3-dihydro-
2-methylnaphthalene   >   3-methylnaphthalene
(1,2,2,3)   (2,2,3,3)

7-Ethyl-2-fluoro-3-nitro-   2-Ethyl-8-fluoro-5-nitro-
cyclooct-1-en-4-yne   >   cyclooct-1-en-3-yne
(1,2,3,4,7)   (1,2,3,5,8)

n) lowest locant for the first named prefix,
    3-Chloro-4-nitroquinoline >   4-Chloro-3-nitroquinoline

6.3
Treatment of the Most Senior Characteristic Group in the Light of the Two Preceding Paragraphs

As has already been settled in the preceding paragraphs that parent component (chain or ring) always has to be chosen as the basis for name-giving which carries the largest number of most senior groups. This axiom is to be taken for granted in the widest sense, for example also when two chains are separated by a ring. If the most senior functional group (characteristic group) is found in the chain, even the largest cyclic substituent cannot prevent the systematic name from being based on that chain. Accordingly, even the longest chain, in selecting the basis for the name, cannot prevail over even the smallest ring if that ring bears the highest ranked functional group.

If the most senior group appears both in the cyclic and in the catenic parent component, that unit serves a reference structure which carries the largest number of these groups. If, however, the number of highest ranked groups is identical for both components the instructions of Section 1.2.1.2.5 must be consulted, according to which that component is preferred which exhibits the highest degree of substitution or the largest number of atoms. Chem. Abstr. in such situations proceeds more consistently insofar as rings are always awarded higher priority than chains.

6.4
Numbering

As a consequence of the presence of functional groups, the numbering rules for parent systems must also be supplemented in a number of essential points. As far as the numbering regulations of Sections 1.1 and 1.2 still leave a choice, subsequent assignment of lowest locants is decided according to the following criteria examined sequentially:

a) indicated hydrogen (even if not denoted explicitly),
b) free valence,
c) highest ranked characteristic group expressed in suffix form,

4-Aminocyclohex-2-en-1-ol   8-Carboxy-4H-fluoren-5-yl

d) multiple bonds altogether, then double bonds,

3,4-Dichlorocyclohex-1-ene   3,7-Dinitrocyclooct-1-en-5-yne

e) lowest locants for all prefix substituents, (detachable) hydro prefixes, "ene"and "yne"positions considered together,

8-Hydroxy-4,5-dimethylazulene-2-carbo-
xylic acid (4.5.8)
not: 4-Hydroxy-7,8-dimethylazulene-2-
carboxylic acid (4.7.8)

f) lowest locant for the (alphabetically) first cited prefix substituent.

1-Methyl-4-nitronaphthalene   but   1-Nitro-4-propylnaphthalene

6.5
Order of Prefixes

As follows from the preceding chapters, prefixes of the most diverse mean-
ing play an often decisive role in the systematization of compound names.
It is therefore necessary to regulate more or less strictly the ordering of
these prefixes in constructing a systematic name.

All prefixes encoding explicit statements about the structure of a pertinent acyclic or cyclic parent system are incorporated into the stem terms,
i.e. treated as non-detachable parts of a name.

Non-detachable prefixes for parent systems:

a) ring-forming: cyclo, bicyclo etc., spiro etc.,
b) ring-cleaving: seco (see steroid nomenclature, Table 22),
c) size-changing: nor, homo (Table 22),
d) fusing: benzo, cycloocta, imidazo etc.,
e) "a" terms of replacement nomenclature: oxa, phospha, azonia, etc.,
f) isomerizing: iso, *sec*, *tert*,
g) hydrogen-indicating,
h) bridge-forming: etheno, benzeno etc.

In contrast, all **substitutive prefixes** (Tables 6, 8) are treated as **detachable**
and ordered alphabetically.

Hydro and **subtractive prefixes** could traditionally be treated as non-detachable **or** detachable (Chem. Abstr., Beilstein); according to a very recent IUPAC proposal they are generally to be classified as **non-detachable.**

It is important to note that **multiplying prefixes** have no influence on the alphabetical order of prefixes. The names of substituted substituents are alphabetized as a whole; otherwise such substituent groups are subject to the same rules as are applied to parent structures, with two exceptions: a) even high-ranked characteristic groups are expressed as suffixes here and b) the linking position (free valence) has the lowest possible locant within the limitations put forth in Section 6.4. For chain substituents this is traditionally always locant 1.

6-Bromomethyl-4,5-dichloro-3a*H*-inden-7-yl
not: 5-Bromomethyl-6,7-dichloro-7a*H*-inden-4-yl

3-Chloromethyl-4-iodo-1,1-dimethylpent-3-enyl

6.6
Isotopically Modified Compounds

Compounds having a nuclide composition deviating from that occurring in nature are defined as **isotopically modified.** The "non-natural" nuclides most important for organic chemistry are ^{11}C, ^{13}C, ^{14}C, ^{15}N, ^{17}O, ^{18}O, ^{34}S, ^{32}P, and above all the hydrogen isotopes listed in Table 18.

Table 18. Symbols and names of hydrogen isotopes

		^1H	^2H [a]	^3H [a]	H [b]
Atom	H$^{\odot}$	Protium	Deuterium	Tritium	Hydrogen
Cation	H$^{\oplus}$	Proton	Deuteron	Trit(i)on	Hydron
Anion	H:$^{\ominus}$	Protide	Deuteride	Tritide	Hydride

[a] If no other modified nuclides are present the symbols D and T can be used.
[b] For cases of unspecified or natural isotopic composition.

All together five types of isotopically modified compounds can be differentiated, one for **isotopically substituted** (a) and four for **isotopically labelled** species (b–e).

a) An **isotopically substituted** compound is present when **all** molecules of this compound have the xenonuclide(s) in (a) well defined position(s) while for all other positions the natural isotopic composition persists. In the name, the nuclide symbols preceded by their locants are placed in round brackets. The element symbols are ordered alphabetically; higher nuclides rank before lower ones.

$C^2H_2Cl_2$
Dichloro($^2H_2$)methane   or CD_2Cl_2
Dichloro(D_2)methane

C^2H_3CN
($^2H_3$)Acetonitrile   or CD_3CN
(D_3)Acetonitrile

$C_6{}^2H_6$
($^2H_6$)Benzene   or C_6D_6
(D_6)Benzene

($^2H_8$)Tetrahydrofuran   or   (D_8)Tetrahydrofuran

$CH_3CO_2{}^2H$
(O-^2H)Acetic acid   or CH_3CO_2D
(O-D)Acetic acid

1-[(^{15}N)Aminomethyl]cyclopentan-1-(^{18}O)ol

b) A **specifically labelled** compound results if one single (singly or multiply) isotopically substituted compound is formally admixed to the corresponding unmodified starting material. In the formula and the name the nuclide symbols, including their multiplicative subscripts, are enclosed in square brackets, e. g.:

$^{13}CH_2{}^2H_2 + CH_4 \rightarrow$ $[^{13}C]H_2[^2H_2]$; $[^{13}C,^2H_2]$Methane

$(C_6H_5)_3[^{32}P] = O$ $[^{32}P]$Triphenylphosphane oxide

$[1\text{-}^{13}C,2,2\text{-}^3H_2]$Naphthalene-1(2$H$)-one

$[^2H]S\text{-}CH_2\text{-}CH(N[^2H_2])\text{-}CO_2[^2H]$ $RS\text{-}[N,N,O,S\text{-}^2H_4]$Cysteine

or: or:

$DS\text{-}CH_2\text{-}CH(ND_2)\text{-}CO_2D$ $(DL)\text{-}[N,N,O,S\text{-}D_4]$Cysteine

$$CH_3\text{-}CH_2\text{-}O\overset{[^{17}O]}{\underset{\parallel}{-}C}\text{-}[^{17}O]\text{-}CH_3$$ O-Ethyl-^{17}O-methyl$[^{17}O_2]$carbonate

c) A **selectively labelled** compound is obtained when a mixture of isotopically substituted compounds is formally added to the corresponding unmodified species. If the positions but not necessarily the numbers of the xenonuclides are known, the nuclide symbols are placed in square brackets in front of the name and the formula, if need be with locants but without multiplying subscripts.

$^{14}C^2H_4 + CH_3{}^2H + CH^2H_3 + CH_4 \rightarrow [^{14}C,^2H]CH_4;\ [^{14}C,^2H]$Methane
(and any other combination of that type)

If the selectively labelled compound is created through admixture of several **clearly defined isotopically substituted** compounds, this is specified by appropriate subscripts to the nuclide symbols involved.

$^{12}CH_4 + CH_2{}^2H_2 + {}^{12}CH^2H_3 + CH_4 \rightarrow [^{12}C_{1;0;1},^2H_{0;2;3}]CH_4$

$[^{12}C_{1;0;1},^2H_{0;2;3}]$Methane

d) An **unselectively labelled** compound is present when position(s) as well as number(s) of the xenonuclides, again taken accout of in square brackets in front of formula and name, are undefined.

$[^{13}C,^2H]CH_3CHClCH_2CO_2H$ 3-Chloro-$[^{13}C,^2H]$butanoic acid

e) **Isotopically deficient** compounds consist of molecules where the natural abundance of one or more nuclides has been depleted; this is specified by the prefix *def.*

$[def^{13}C]CHCl_3$ $[def^{13}C]$Chloroform

From a reversed viewpoint this of course corresponds to an "**enrichment**"of the natural isotope, thus:

$[^{12}C]CDCl_3$ $[^{12}C]$Chloroform

Table 19 recapitulates once again the main statements of this paragraph in the form of a synopsis.

Table 19. Formulae and names of isotopically modified compounds in direct comparison

Modification	Formula	Name
Unmodified	CH_3CH_2OH	Ethanol
Isotopically substitued	$C^2H_3CH_2-{}^{18}O^2H$	$(2,2,2-{}^2H_3)Ethan({}^2H,{}^{18}O)ol$
		or: $(O,2,2,2-{}^2H_4,{}^{18}O)Ethanol$
Specifically labelled	$[{}^{13}C]H_3-C[{}^2H_2]-O[{}^2H]$	$[2-{}^{13}C;1,1-{}^2H_2]Ethan[{}^2H]ol$
		or: $[2-{}^{13}C;O,1,1-{}^2H_3]Ethanol$
Selectively labelled	$[O,2-{}^2H]CH_3CH_2OH$	$[O,2-{}^2H]Ethanol$
	$[2-{}^2H_{2;2},{}^{18}O_{0;1}]CH_3CH_2OH$	$[2-{}^2H_{2;2},{}^{18}O_{0;1}]Ethanol$
Unselectively labelled	$[{}^2H]CH_3CH_2OH$	$[{}^2H]Ethanol$ oder $[D]Ethanol$
Isotopically deficient	$CH_3-[def\,{}^{13}C]H_2-OH$	$[1-def\,{}^{13}C]Ethanol$
or	$CH_3-[{}^{12}C]H_2-OH$	$[1-{}^{12}C]Ethanol$

6.7
Specifications of Stereochemistry

All constitutive elements of a compound having been duly considered in compliance with the existing rules, still another set of descriptors has to be envisaged to account properly for all stereochemical features involved. Definitive status has already been attained by the rules for characterizing *cis/trans* isomers, stereogenic (formerly: asymmetric) centers, and *exo/endo* relationships in bicyclic compounds, as will be outlined in subsequent sections. A unified procedure for the treatment of other stereogenic units will be given at the end of this chapter.

6.7.1
cis/trans Isomerism; the *E/Z* Convention

cis or *trans*-arrangement of atoms or groups obtains when these are found at the same or the opposite side of a reference plane common to both stereoisomers of the respective molecule. For double bond systems this is the plane containing the π-bond, for cyclic systems it is that plane to which the ring skeleton approximates most closely.

6.7.1.1
Double Bond Systems

For vicinally disubstituted olefins (and similar double-bond systems) the traditional prefixes *cis* and *trans* can be retained, although for all higher substituted systems the descriptors (*Z*) and (*E*) as defined on the basis of the **CIP sequence rules** must be applied.

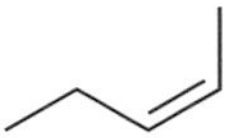

cis-Pent-2-ene
(*Z*)-Pent-2-ene

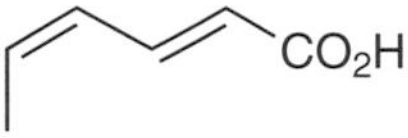

2-*trans*,4-*cis*-Hexa-2,4-dienoic acid
(2*E*, 4*Z*)-Hexa-2,4-dienoic acid

If in the above example the **conformation** relative to the 3,4-single bond should also be specified, the supplemental prefixes *s-cis* (vs. *s-trans*) will serve that purpose, thus giving: *s-trans*,(2E,4Z)-hexadienoic acid.

1-*cis*,3-*trans*-Cycloocta-1,3-diene
(1*Z*, 3*E*)-Cycloocta-1,3-diene

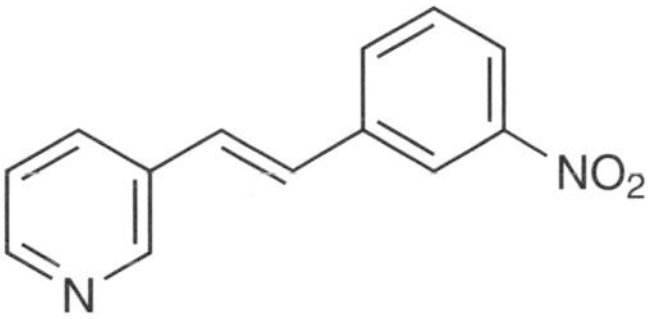

3-[*trans*- or (*E*)-2-(3-Nitro-phenyl)ethenyl]pyridine

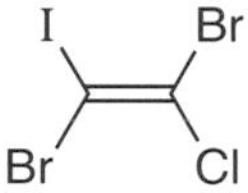

(*Z*)-1,2-Dibromo-1-chloro-2-iodoethene

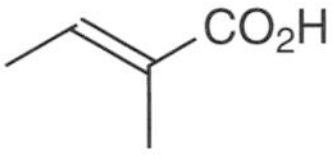

(*E*)-2-Methylbut-2-enoic acid

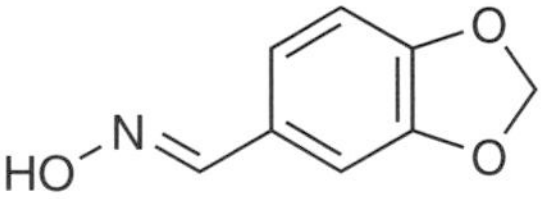

Piperonal (*E*)-oxime
trad.: Piperonal *syn*-oxime

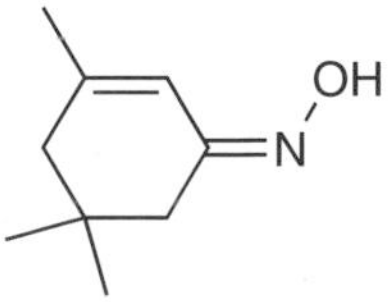

3,5,5-Trimethylcyclohex-2-enone (*Z*)-oxime

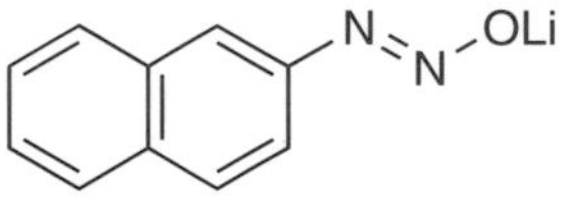

Lithium naphthalen-2-yl-(*E*)-diazenolate
trad.: Lithium naphthalene-2-*anti*-diazoate

6.7.1.2
Ring Systems

For bicyclic systems and non-geminally disubstituted monocycles the traditional *cis/trans* descriptors are used exclusively.

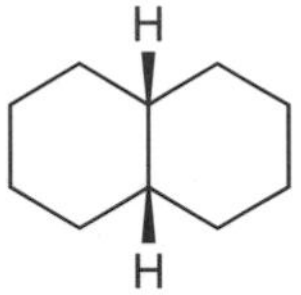

cis-Decahydronaphthalene
(*cis*-Decalin)

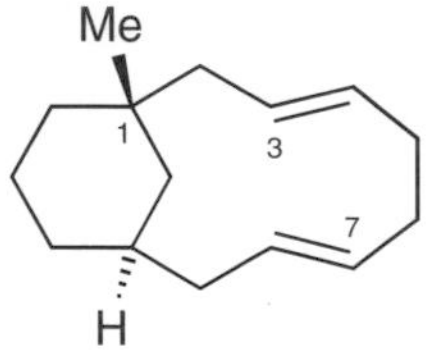

1-Methyl-*trans*-bicyclo[8.3.1]tetradeca-
3-*trans*,7-*trans*-diene

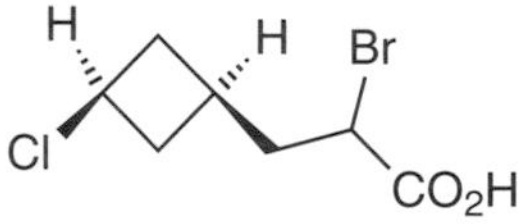

conj.: α-Bromo-*cis*-3-chlorocyclobutane-
propanoic acid
subst.: 2-Bromo-3-(*cis*-3-chlorocyclobutyl)
propanoic acid

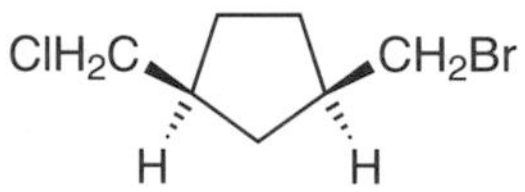

trans-1-(Bromomethyl)-2-
(chloromethyl)cyclopentane

For rings bearing more than two substituents attached to saturated ring members an extended *cis/trans* formalism must be applied. Here, the highest ranked suffix-substituent or, if there is none, the most senior prefix-substituent of the lowest numbered doubly substituted ring member serves as reference group. This reference group is now labelled with a lower case *r* in front of the respective locant; subsequently the most senior substituents at the other ring atoms are referred to the *r*-substituent by the lower case letters *c*(is) or *t*(rans) placed again in front of the corresponding locants. All rankings here must of course be done on the basis of the CIP sequence rules (see Section 6.7.2.1).

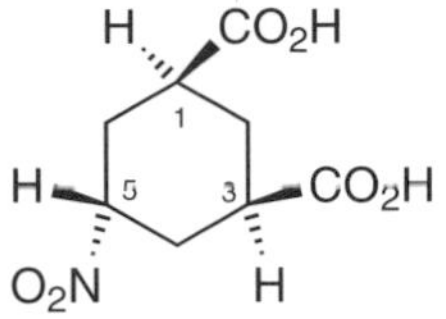

t-5-Nitrocyclohexane-r-1,c-3-dicarboxylic acid

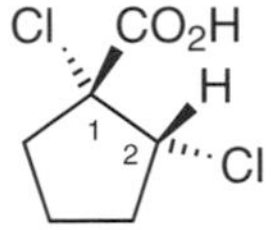

1,t-2-Dichlorocyclopentane-r-1-carboxylic acid

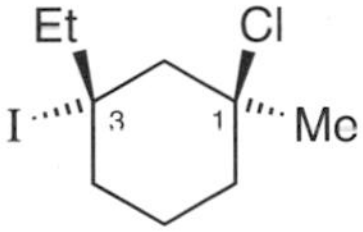

r-1-Chloro-3-ethyl-1-methyl-t-3-iodocyclohexane

All the above examples contain stereogenic centers and can therefore also be characterized according to the rules for the specification of absolute configuration as described in the following paragraphs.

6.7.2
Specification of Absolute and Relative Configuration

6.7.2.1
Compounds with Stereogenic (Traditionally Asymmetric) Carbon (and Analogous) Centers

Chiral molecules with stereogenic centers of known absolute configuration are differentiated by the stereo descriptors R and S assigned according to the sequence rules stated below. For tetracoordinated centers these descriptors are obtained by placing the **lowest ranked ligand** (atom or substituent group) away from the viewer so that the remaining three ligands exhibit a clockwise (R) or anticlockwise (S) arrangement, depending on their ranking[1]. Such steric arrangements can be characterized by the collective term **tripodal stereogenic unit.**

---

[1] For amino acids, carbohydrates, and related compound types the traditional descriptors D and L are preferred.

Priority sequence of ligands according to the **CIP rules**[2].

1. Higher > lower atomic number
2. Higher > lower mass number
3. *Z; cis*   > *E; trans*
4. *R,R; S,S > R,S; S,R. r > s*
5. *R,M*   > *S,P*

For most of the practically important cases the first three criteria are sufficient for assigning conclusive priorities.

To visualize the stereochemical relationships in tetrahedral molecules the following graphical conventions have been established, among which can be freely chosen according to personal gusto.

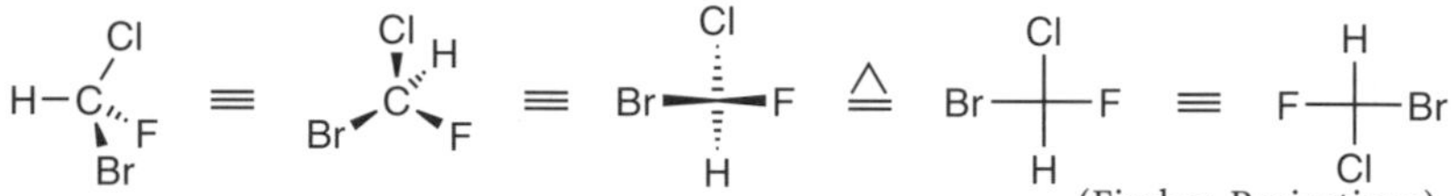

(*R*)-Bromochlorofluoromethane (Br > Cl > F > H)   (Fischer-Projections)

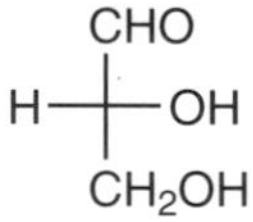

HO > CHO > CH_2OH > H

D-Glyceraldehyde
(*R*)-2,3-Dihydroxypropanal

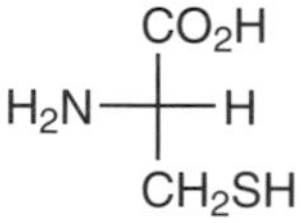

L-Cysteine
(*R*)-2-Amino-3-sulfanylpropanoic acid

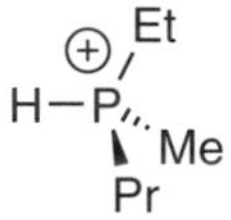

(*R*)-Ethylmethylpropylphosphonium

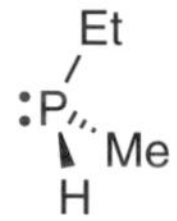

(*R*)-Ethylmethylphosphane (H > :)

---

[2] First comprehensive publication: R.S. Cahn, C.K. Ingold, V. Prelog, Angew. Chem. **1966**, *78*, 413. Substantially modified and expanded version: V. Prelog, G. Helmchen, Angew. Chem. **1982**, *94*, 614; Intern. Edition, **1982**, *21*, 567.

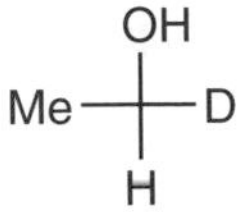

(S)-(1-D₁)Ethanol — $(S)\text{-}(1\text{-}D_1)$Ethanol

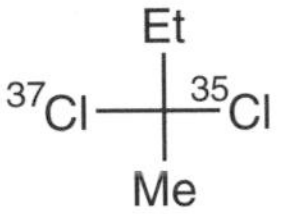

(S)-2,2-Di(^{37}Cl, ^{35}Cl) chlorobutane

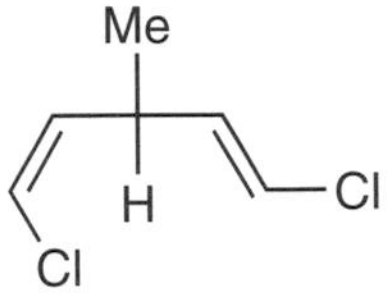

(S)-1,5-Dichloro-3-methylpenta-1-*cis*,4-*trans*-diene

When several stereogenic (formerly: asymmetric or chiral) centers are present in a molecule the over-all configuration is characterized by a **set of R,S symbols** preceding the compound name.

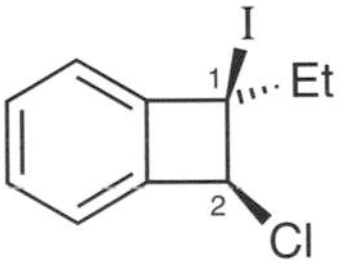

$(1R,2S)$-2-Chloro-1-ethyl-1,2-dihydro-1-iodocyclobutabenzene

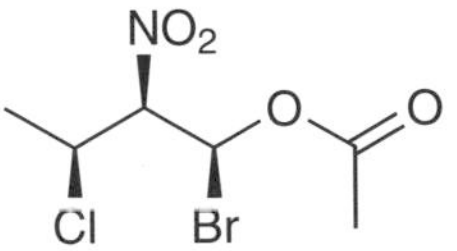

$(1S,2R,3S)$-1-Bromo-3-chloro-2-methyl-2-nitrobutyl acetate

If only **relative configurations** should be marked the descriptors R^*/S^* are applied for both enantiomers in such a way that the center with the lowest locant is arbitrarily labelled with R^*. Alternatively the unstarred symbols R/S can be utilized here too when specified by the prefix *rel*.

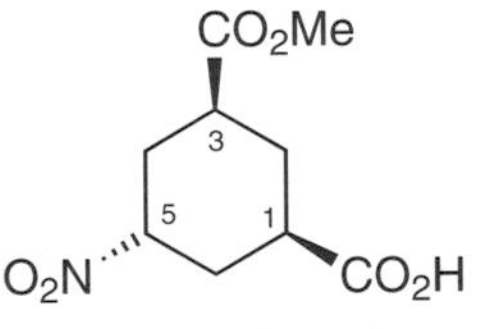

or

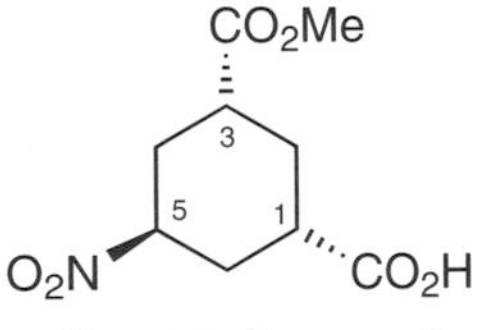

$(1R^*,3S^*,5S^*)$- oder *rel*-$(1R,3S,5S)$-3-Methoxycarbonyl-5-nitrocyclo-hexane-1-carboxylic acid

For compounds derivable (if only in the widest sense) from carbohydrates and containing only two stereogenic centers the traditional descriptors **erythro/threo** are still frequently used and, as indicators of **relative ste-**

reochemistry, obtain each for both enantiomers of the **racemic pair** (one of two) under scrutiny.

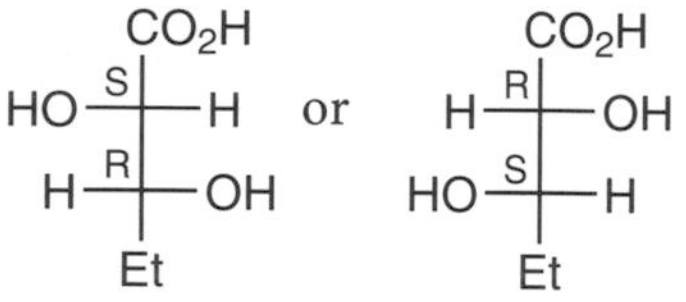

($2R^*,3S^*$)-2,3-Dihydroxypentanoic acid
threo-2,3-Dihydroxypentanoic acid

The following two dicarboxylic acids represent achiral **diastereomeric meso forms** with a **pseudostereogenic** (formerly: pseudoasymmetric) C-atom, C-5, which is given the descriptor *r* or *s*, respectively.

(1R,3S,5s)-   (1R,3S,5r)-
5-Nitrocyclohexane-1,3-dicarboxylic acid

Although for polycyclic systems too all stereogenic centers can be unambiguously characterized as shown above, for bicyclo[x.y.z]alkanes with $x \geqq y > z > 0$ a specific *endo/exo,syn/anti* formalism is generally preferred. Thus, to a group pointing towards the bridge with the highest locants is assigned the descriptor *exo*, if pointing away the descriptor *endo*; at the same time a group located at that same bridge and pointing toward the bridge with the lowest locants is characterized by the prefix *syn*, if pointing away by *anti*.

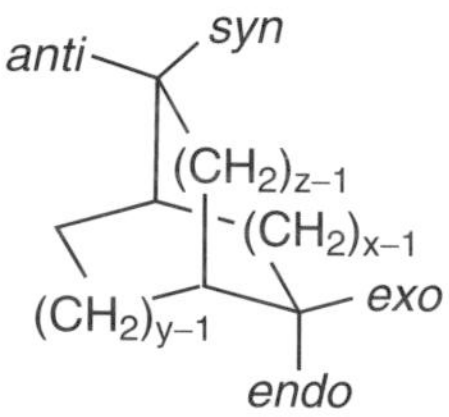

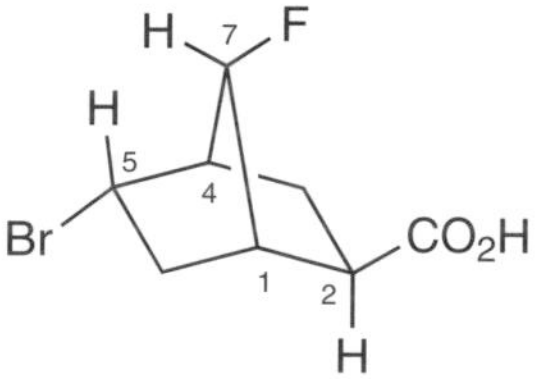

5-*endo*-Bromo-7-*syn*-fluorobicyclo[2.2.1]heptane-2-*exo*-carboxylic acid
or: (1S,2R,4R,5S,7R)-5-Bromo-7-...

6.7.2.2
Molecules with Helical Stereogenic Units

Stereogenic units of quite different types are responsible for the existence of enantiomers in the case of **axially** and **planary chiral** molecules whose **helicities** – P(lus) or M(inus) – can most easily be derived with the help of appropriate **four-point figures** inscribed in the respective structural formulae.

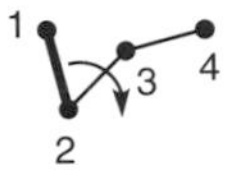

The appertaining stereodescriptors depend on whether the **syn-periplanar** arrangement of connecting lines 1 – 2 and 3 – 4 (corresponding to a superposition of planes 1, 2, 3 and 2, 3, 4) requires a right (P) or left (M) turn around axis 2 – 3.

6.7.2.2.1
Screw-like Molecules (One Chirality Axis)

Screw-like structures are found above all for molecules such as helicenes, 1,3-disubstituted allenes, 2,2′-disubstituted biphenyls, etc., displaying C_2 symmetry. With the aid of the 4-point figures indicated in the formulae the absolute helicities (P or M) can straightforwardly be ascertained. The currently still customary descriptors aS (or S_a) and aR (or R_a) regrettably correspond only **inversly** to the much more generally applicable **helicity symbols** P and M.

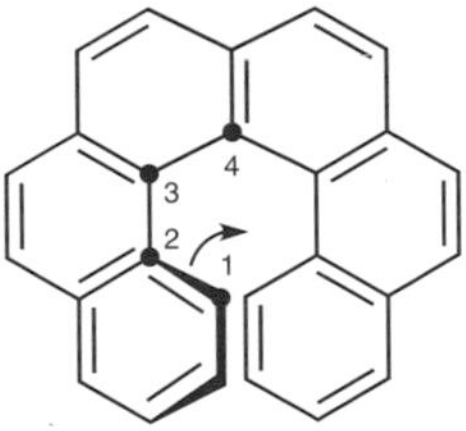

(P)-Hexahelicene

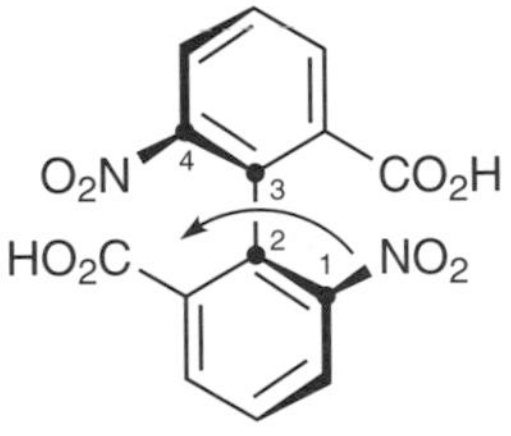

(M)-6,6′-Dinitrobiphenyl-2,2′-dicarboxylic acid

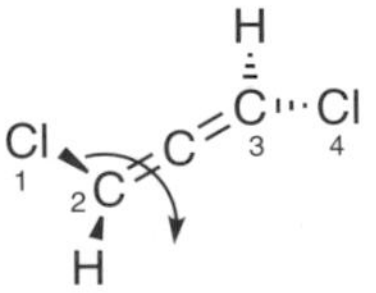

(P)-1,3-Dichloropropadiene

6.7.2.2.2
Propeller-like Molecules (Several Chirality Axes)

Structures of symmetry D_n can be viewed as **n-bladed propellers** where, in a way, n screw fragments are grouped around a real or fictitious central unit. This category includes above all triaryl-element and comparable polyaryl-derivatives, doubly bridged biphenyls, and a number of twisted molecules. Here again the apparent helicities can easily be determined by evaluating the 4-point figures imprinted into the structural formulae.

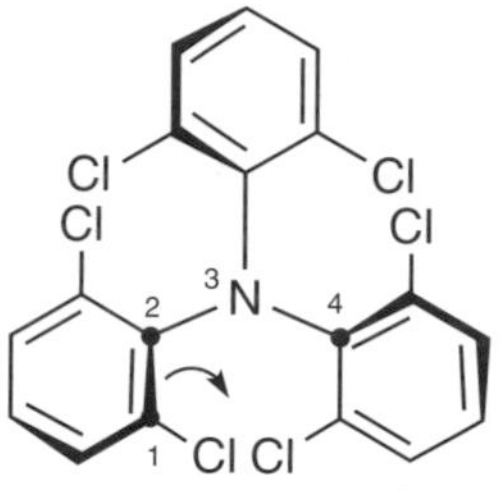

$(P)^3$-Tris(2,6-dichlorophenyl)azane
Chem. Abstr.:
(P)-N,N-Bis(2,6-dichlorophenyl)-2,6-di-chlorobenzenamine (or: ... aniline)

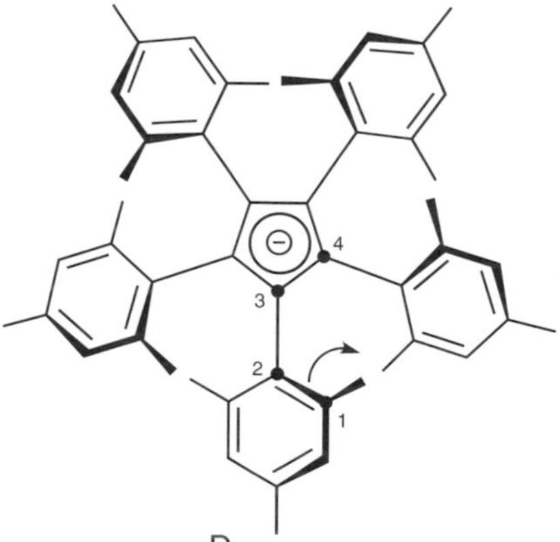

(P)- or more precisely (P,P,P,P,P)-Pentamesitylcyclopentadienide

---

[3] This helicity descriptor should actually read (P,P,P) because in principle (though never observed as yet) one phenyl group could also be rotated **in the contrary sense,** thus leading to a (P,P,M) diastereomer.

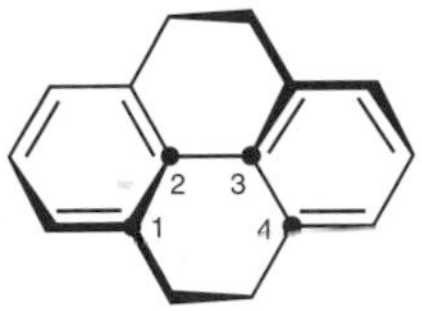

(P)-4,5,9,10-Tetrahydropyrene

D_2

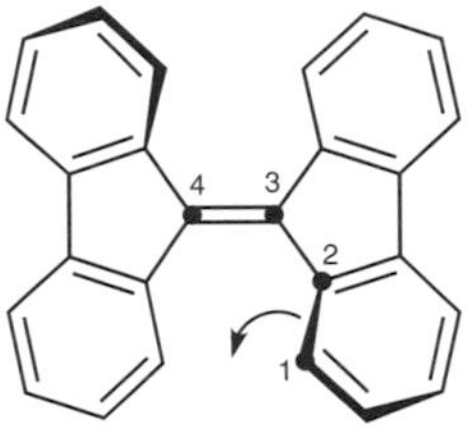

(M)-Bi(fluoren-9,9'-ylidene)

D_2

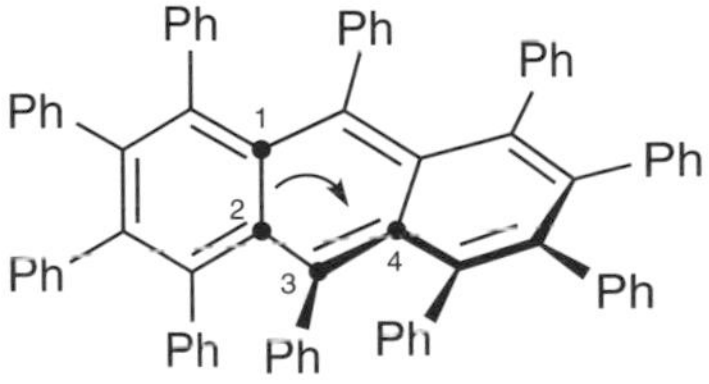

(M)-Decaphenylnaphthalene

D_2

Tris-chelated compounds with three symmetrical bidendate ligands (in inorganic chemistry these are the prototypical tris-chelate complexes of transition metals) can also be conceived as three-bladed propellers with their helicities specified accordingly.

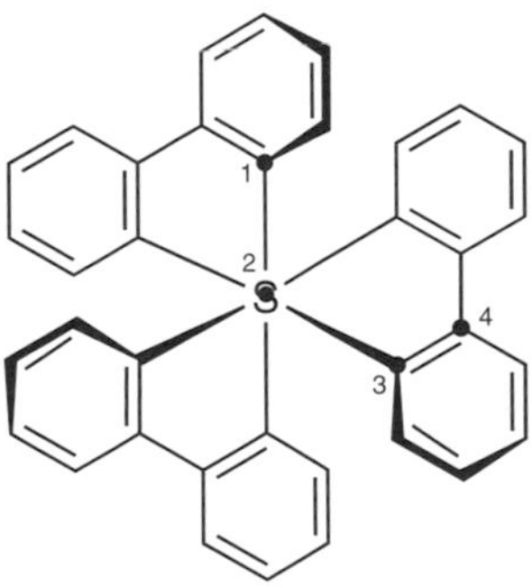

(M)[4]-Tris(biphenyl-2,2'-diyl)-λ^6-sulfane

---

[4] Inorganic chemists prefer here the symbols Λ, λ (lamda) and Δ, δ (delta).

6.7.2.3
Molecules Exhibiting Planar Chirality

All the subsequently shown compounds possess a chirality plane and their helicities can again be described with respect to interposed 4-point figures. In this class belong in the first place so-called *ansa*-compounds, unsymmetrically substituted cyclophanes, and certain cycloolefin conformations.

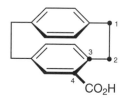

Chem. Abstr.: (*P*)-12-Methyl-14-nitrobicyclo[9.2.2]pentadeca-1(13),11,14-triene
New **IUPAC**-cyclophane name:
(*P*)-1²-Methyl-1⁶-nitro-1(1,4)-benzenacylodecaphane

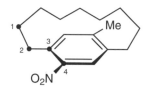

(*M*)-Tricyclo[8.2.2.2^{4,7}]hexadeca-1(12),4,6,10,13,15-hexaene-5-carboxylic acid

(*M*)-1,4(1,4)-Dibenzenacyclohexaphane-1²-carboxylic acid

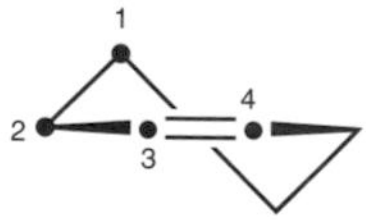

(*M*)-Half chair cyclohexene

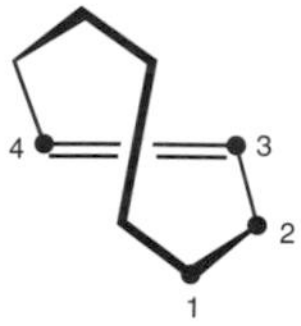

(*M*)-Crown cyclooctene

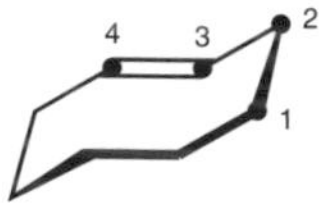

(*P*)-Chair cyclooctene

6.7.3
Concluding Remarks

The author is well aware of the fact that many chemists still advocate use of the traditional R/S symbols also for at least some, if not all, compound types with helical stereogenic units, even if these are **not in concordance** with the proper helicity descriptors as in the case of axially chiral compounds where $P \neq R_a \triangleq M$ and $M \neq S_a \triangleq P$. For the sake of consistency, though, it appears highly advisable to discern the clearly distinct modes of stereogenicity – tripodal on the one hand versus helical (4-point figure) on the other – not only at the conceptual level but also with respect to the appertaining descriptors, hic R/S illic P/M.

7 Appendix

7.1
Complete List of "a"-terms Utilized in Replacement and Heterane Nomenclature

Table 20. Complete list of replacement terms ("a" terms) in descending order of priority

Element	"a" Prefix	Element	"a" Prefix	Element	"a" Prefix
F	Fluora	Au	Aura	Eu	Europa
Cl	Chlora	Ni	Nickela	Gd	Gadolina
Br	Broma	Pd	Pallada	Tb	Terba
I	Ioda	Pt	Platina	Dy	Dysprosa
At	Astata	Co	Cobalta	Ho	Holma
O	Oxa	Rh	Rhoda	Er	Erba
S	Thia	Ir	Irida	Tm	Thula
Se	Selena	Fe	Ferra	Yb	Ytterba
Te	Tellura	Ru	Ruthena	Lu	Luteta
Po	Polona	Os	Osma	Ac	Actina
N	Aza	Mn	Mangana	Th	Thora
P	Phospha	Tc	Techneta	Pa	Protactina
As	Arsa	Re	Rhena	U	Urana
Sb	Stiba	Cr	Chroma	Np	Neptuna
Bi	Bisma	Mo	Molybda	Pu	Plutona
C	Carba	W	Tungsta[a]	Am	America
Si	Sila	V	Vanada	Cm	Cura
Ge	Germa	Nb	Nioba	Bk	Berkela
Sn	Stanna	Ta	Tantala	Cf	Californa
Pb	Plumba	Ti	Titana	Es	Einsteina
B	Bora	Zr	Zircona	Fm	Ferma
Al	Alumina	Hf	Hafna	Md	Mendeleva
Ga	Galla	Sc	Scanda	No	Nobela
In	Inda	Y	Yttra	Lr	Lawrenca
Tl	Thalla	La	Lanthana	Be	Berylla
Zn	Zinca	Ce	Cera	Mg	Magnesa
Cd	Cadma	Pr	Praseodyma	Ca	Calca
Hg	Mercura	Nd	Neodyma	Sr	Stronta
Cu	Cupra	Pm	Prometha	Ba	Bara
Ag	Argenta	Sm	Samara	Ra	Rada

[a] Also Wolframa.

7.2
Tables of Customary Trivial (and Semitrivial) Names

Extended series of trivial names that are to be retained (or just too frequently used to be simply omitted) have already been compiled in various tables of Chapter 1 of this book (fused polycycles: Table 1, p. 16, heterocycles: Tables 2 and 3, p. 44, 50). For the numerous trivial names to be dealt with in the domain of the functional compound classes of Chapter 2 it has been found more appropriate not to include them directly in the text but to confine them to a tabular appendix. Specific hydrocarbon systems such as terpenes and steroids, whose nomenclature is widely dominated by trivial names, are also accounted for in this appendix.

Table 21. Cyclic terpene hydrocarbons[a]

p-Menthane[b]			Thujane[c]			Carane[d]

[a] Acyclic terpenes and tetramethylcyclohexane-type representatives are named systematically. Unsaturated derivatives of the above trivially named systems are characterized in the usual way, e.g.: 3-p-menthene, 2,5-norbornadiene etc.

[b] If further alkyl groups are present systematic names are used.

[c] If additional methyl or isopropyl side chains (or more than one methylene groups) are present naming is done systematically. Naming of other substitution products is again related to the respective trivial names.

[d] If additional side chains (other than methyl or methylene groups) are present naming is based on the respective trivial names. Other substitution products of these skeletons are treated as **norcarane, norpinane**, and **norbornane** derivatives.

Table 21 (continued)

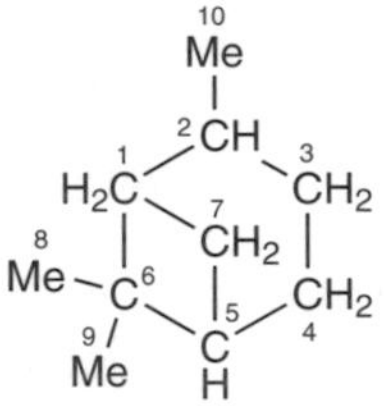

Pinane[d]	Bornane[d]	Norcarane
Norpinane	Norbornane	Camphene[b]

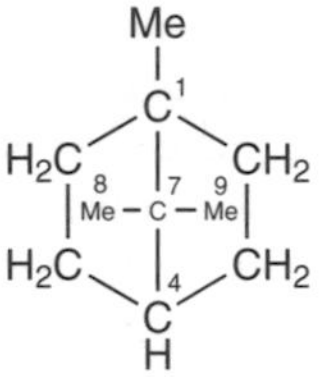

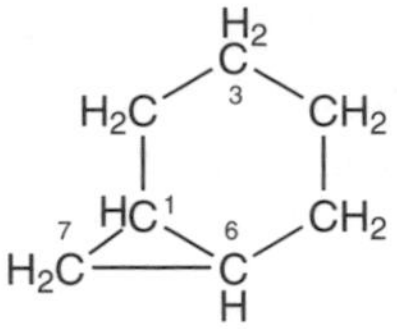

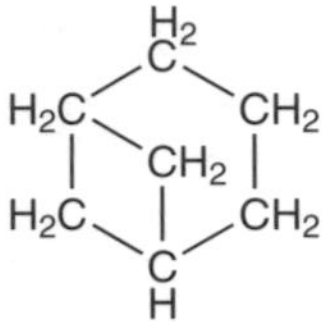

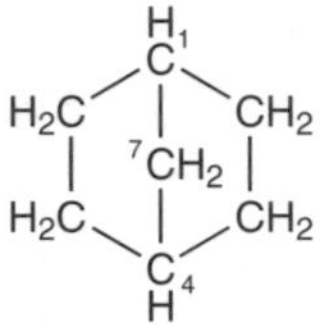

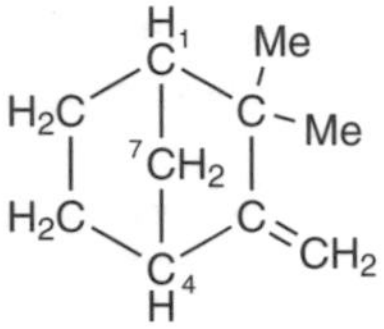

Table 22. Steroids

Details of the very specific steroid nomenclature must be taken from the original rule codex (Pure Appl. Chem. **1989**, *61*, 1783; see also l.c.[9] on page 5). Here only the fundamental compound types and the most elementary rules can be considered.

Fundamental types

The following scheme displays the (exceptional) numbering conventions that hold also for the fully unsaturated (mancude) basic skeleton, namely, 15H-cyclopenta[a]phenanthrene.

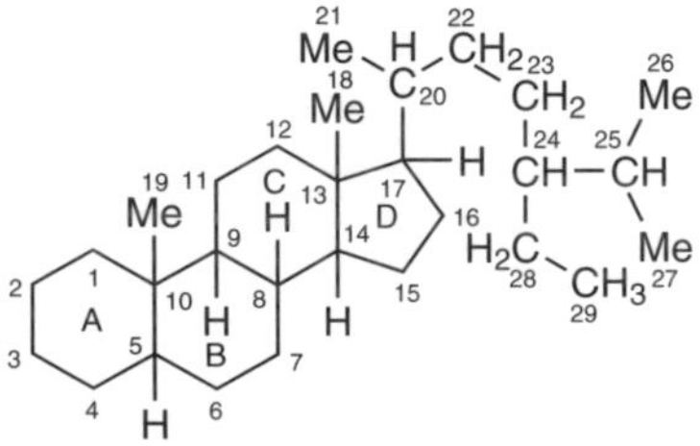

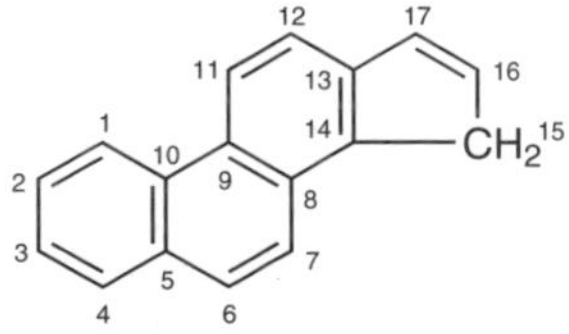

15H-Cyclopenta[a]phenanthrene

Table 22 (continued)

Groups protruding from the projection plane have their bonds in bold print and the designator β; those receding behind this plane are shown with broken bonds and the descriptor α.

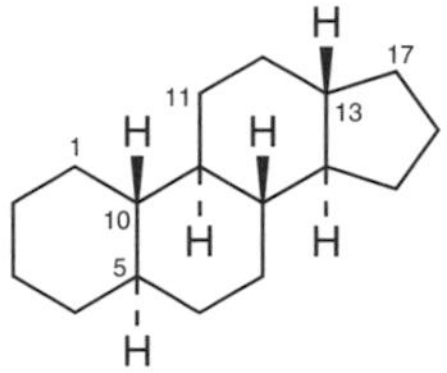

5α-Gonan 5β-Estrane

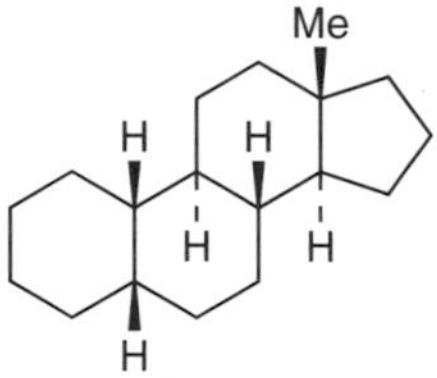

Undulated bonds and the designator ζ are used when it is not intended or possible to specify the actual steric orientation of a group.

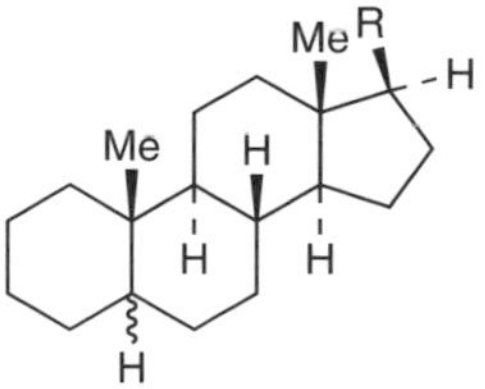

R	Trivial name
H	5ζ-Androstane
$\overset{20\ 21}{C_2H_5}$	5ζ-Pregnane
$\overset{20}{CH}(CH_3)CH_2CH_2CH_3$	5ζ-Cholane[a]
$\overset{20}{CH}(CH_3)CH_2CH_2CH_2CH(CH_3)_2$	5ζ-Cholestane[a]
$\overset{20}{CH}(CH_3)CH_2CH_2\overset{24}{CH}(CH_3)CH(CH_3)_2$	5ζ-Ergostane[a,b]
$\overset{20}{CH}(CH_3)CH_2CH_2\overset{24}{CH}(C_2H_5)CH(CH_3)_2$	5ζ-Stigmastane[a,c]

[a] 20R, [b] 24S, [c] 24R configuration.

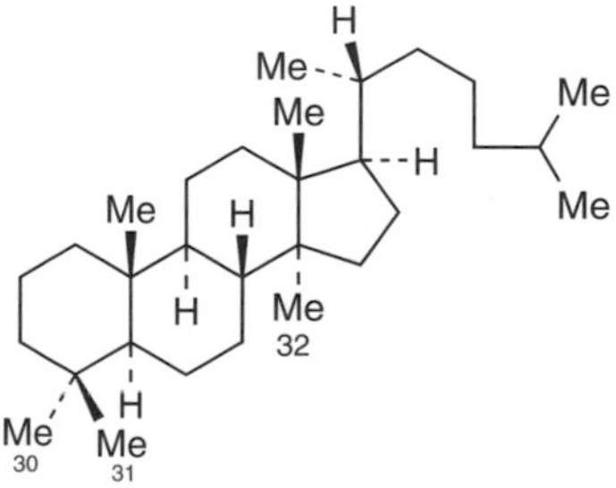

5α-Lanostane 5α-Protostane

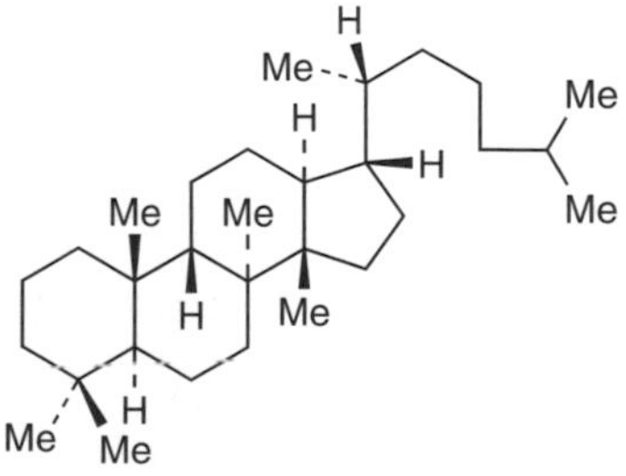

Table 22 (continued)

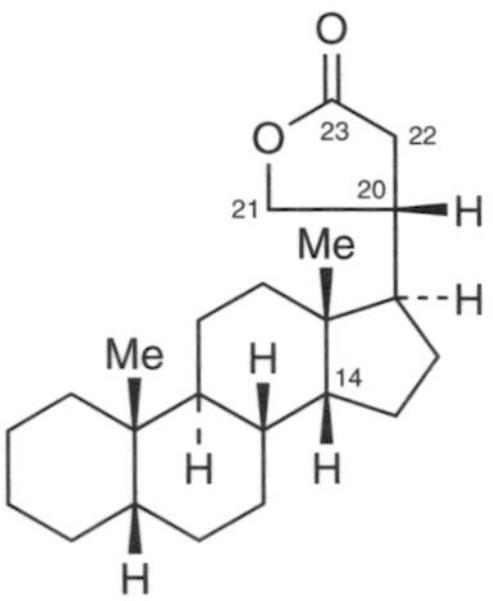

5β,14β-Cardanolide

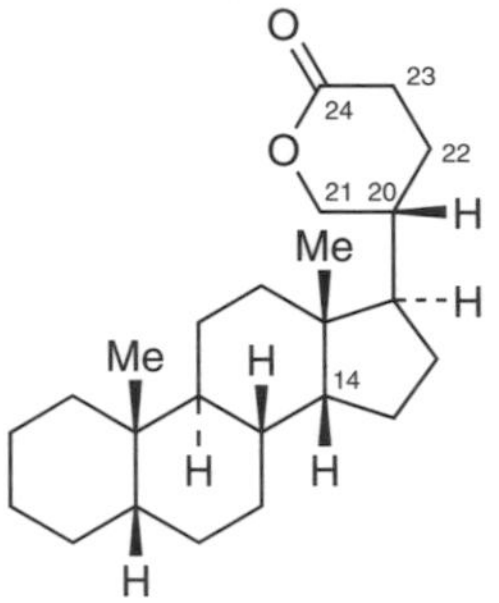

5β,14β-Bufanolide

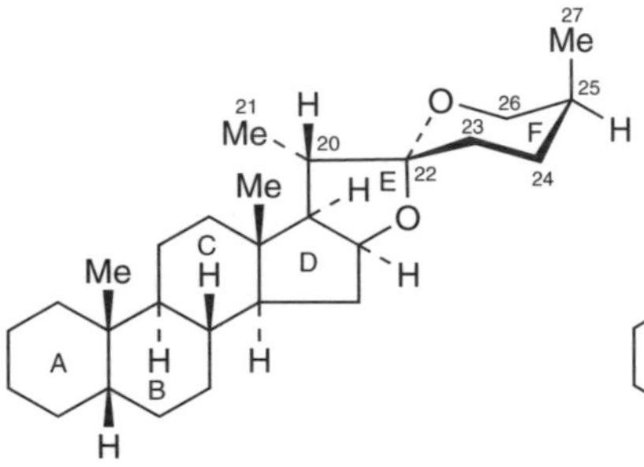

(25S)-5β-Spirostan

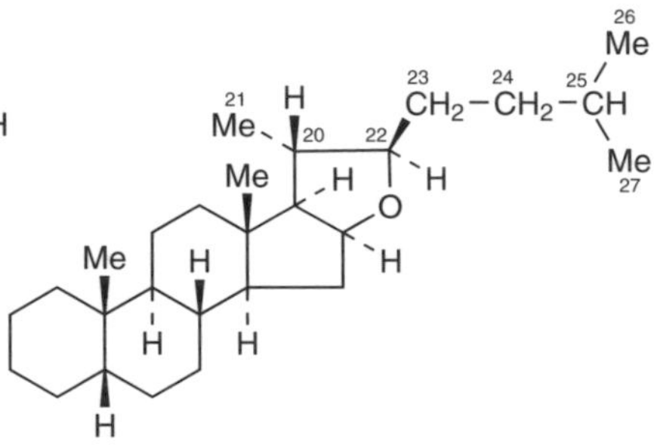

(22R)-5β-Furostan

Unsaturated and substituted derivatives of these fundamental types are treated according to the general proceedings of organic chemical nomenclature. Several important examples for which, in addition, regular trivial names are retained are shown below.

Trivial name	Semisystematic name
Aldosterone	18,11-Hemiacetal of 11β,21-Dihydroxy-3,20-dioxo-4-pregnen-18-al
Androsterone	3α-Hydroxy-5α-androstan-17-one
Cholecalciferol	9,10-Seco-5,7,10(19)-cholestatrien-3β-ol
Cholesterol	5-Cholesten-3β-ol
Cholic acid	3α,7α,12α-Trihydroxy-5β-cholan-24-oic acid
Corticosterone	11β,21-Dihydroxy-4-pregnene-3,20-dione
Cortisol	11β,17,21-Trihydroxy-4-pregnene-3,20-dione
Cortisone	17,21-Dihydroxy-4-pregnene-3,11,20-trione
Deoxycorticosterone	21-Hydroxy-4-pregnene-3,20-dione
Ergocalciferol	9,10-Seco-5,7,10(19),22-ergostatetraen-3β-ol
Ergosterol	5,7,22-Ergostatrien-3β-ol
Estradiol-17α	1,3,5(10)-Estratrien-3,17α-diol
Estriol	1,3,5(10)-Estratriene-3,16α,17β-triol

Table 22 (continued)

Trivial name	Semisystematic name
Estrone	3-Hydroxy-1,3,5(10)-estratrien-17-one
Lanosterol	8,24-Lanostadien-3β-ol
Lithocholic acid	3α-Hydroxy-5β-cholanoic-24-acid
Progesterone	4-Pregnene-3,20-dione
Testosterone	17β-Hydroxy-4-androsten-3-one

Modified types

Additional linking of two C-atoms is characterized by the prefix **cyclo** placed after the pertaining locant and stereo designators.

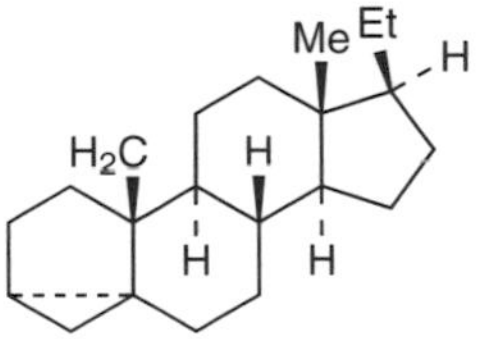

3α,5-Cyclo-5α-pregnane   11β,19-Cyclo-5α-androstane

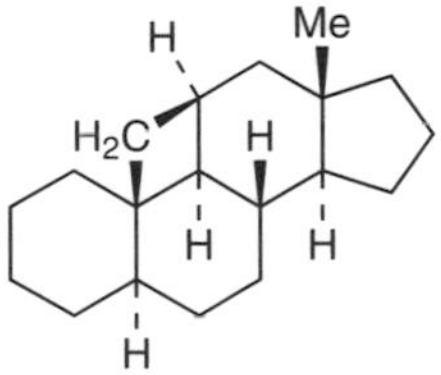

Ring cleavage is indicated accordingly with the prefix **seco**.

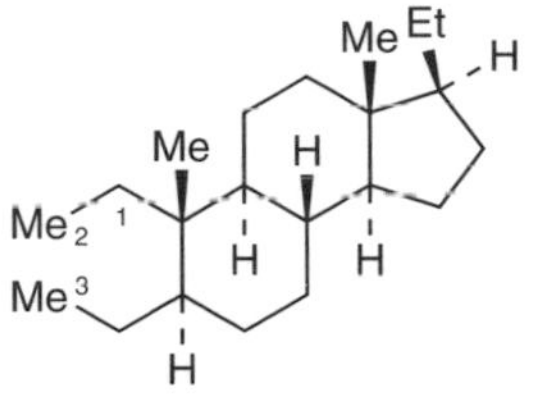

2,3-Seco-5α-pregnane 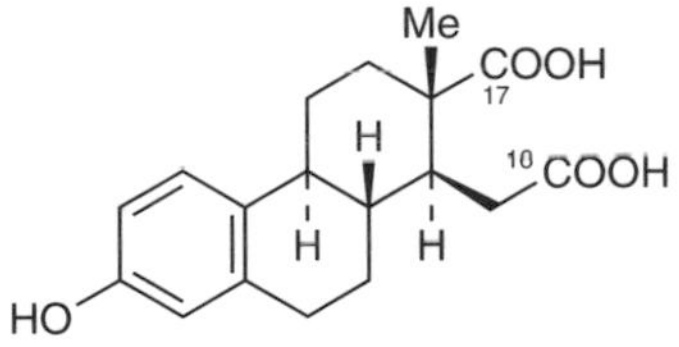   3-Hydroxy-16,17-seco-1,3,5(10)-
                              ö(e)stratriene-16,17-dioic acid

Elimination of a $>$CH$_2$-group from a side chain or of a methyl group from the skeleton is indicated by the prefix **nor** coming behind the locant of the omitted group.

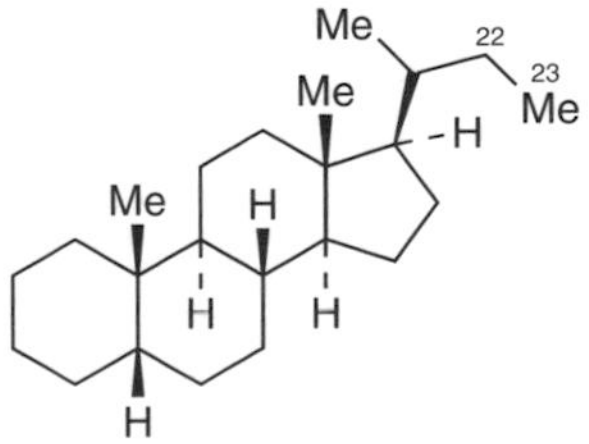

24-Nor-5β-cholane 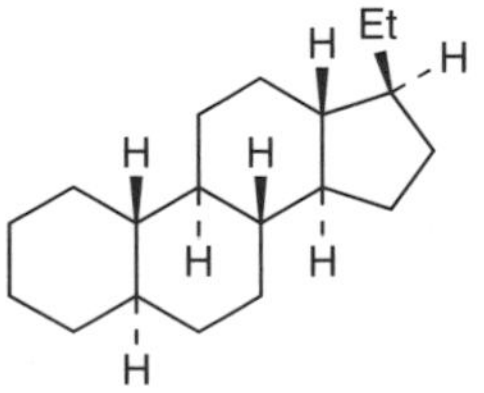   18,19-Dinor-5α-pregnane

Table 22 (continued)

Ring contraction too is indicated by the syllable **nor** while ring expansion is expressed with the prefix **homo**. In both cases the designator of the ring in question is placed in front of the so defind ring modifiers. The original numbering is maintained, but for a **nor** derivative the highest locant of the modified ring disappears while for the **homo** product the highest locant is supplemented by lower case letters a,b ... etc.

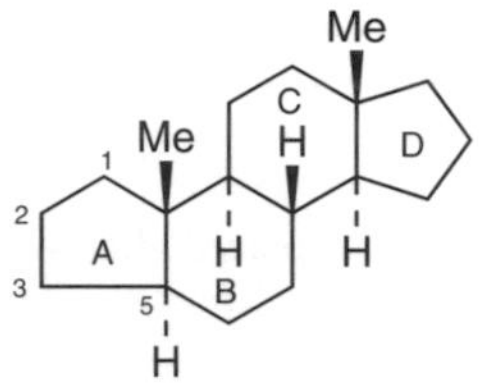

A-Nor-5α-androstane

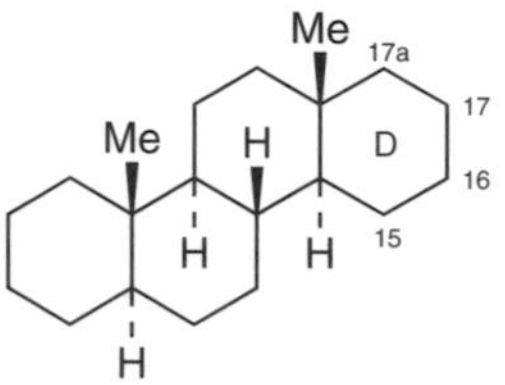

D-Homo-5α-androstane

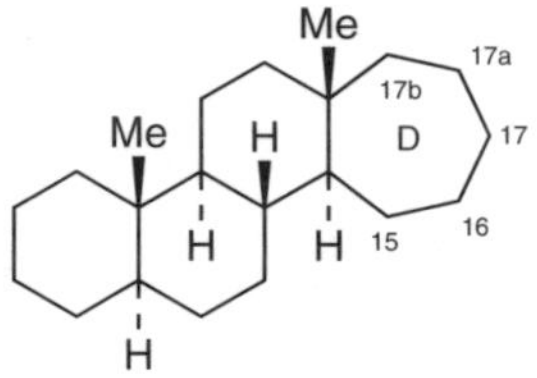

D-Dihomo-5α-androstane

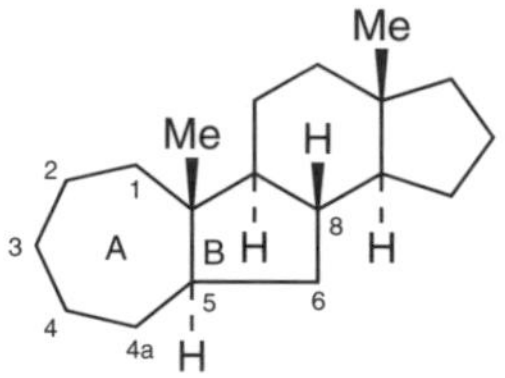

A-Homo-*B*-nor-5α-androstane

Heterocyclic steroid analogues are treated according to the rules of replacement nomenclature (see p. 51).

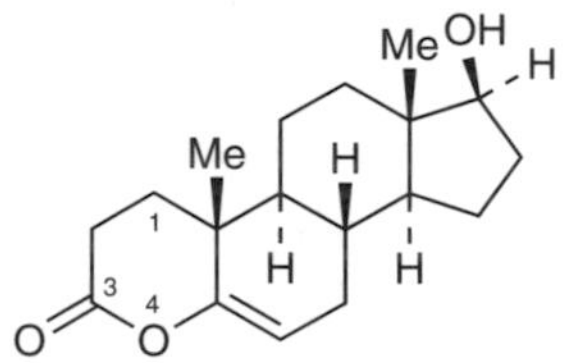

17β-Hydroxy-4-oxa-5-androsten-3-one

Table 23. Carboxylic acids and their substituent groups

Name of the acid[a]	Group name	Formula
Saturated aliphatic carboxylic acids (limiting list)		
Formic acid	Formyl	$HCO-\{$
Acetic acid	Acetyl	$H_3C-CO-\{$
Propionic acid	Propionyl	$H_3C-CH_2-CO-\{$
Butyric acid	Butyryl	$H_3C-(CH_2)_2-CO-\{$
Isobutyric acid	Isobutyryl	$(H_3C)_2CH-CO-\{$
Valeric acid	Valeryl	$H_3C-(CH_2)_3-CO-\{$
Isovaleric acid[a]	Isovaleryl	$(H_3C)_2CH-CH_2-CO-\{$
Pivalic acid[a]	Pivaloyl	$(H_3C)_3C-CO-\{$
Lauric acid	Lauroyl	$H_3C-(CH_2)_{10}-CO-\{$
Myristic acid[a]	Myristoyl	$H_3C-(CH_2)_{12}-CO-\{$
Palmitic acid[a]	Palmitoyl	$H_3C-(CH_2)_{14}-CO-\{$
Stearic acid[a]	Stearoyl	$H_3C-(CH_2)_{16}-CO-\{$

Saturated aliphatic dicarboxylic acids (non-limiting list)		
Oxalic acid	Oxalyl	$\{-CO-CO-\{$
Malonic acid	Malonyl	$\{-CO-CH_2-CO-\{$
Succinic acid	Succinyl	$\{-CO-(CH_2)_2-CO-\{$
Glutaric acid	Glutaryl	$\{-CO-(CH_2)_3-CO-\{$
Adipic acid	Adipoyl	$\{-CO-(CH_2)_4-CO\{$
Pimelic acid	Pimeloyl	$\{-CO-(CH_2)_5-CO-\{$
Suberic acid	Suberoyl	$\{-CO-(CH_2)_6-CO-\{$
Azelaic acid	Azelaoyl	$\{-CO-(CH_2)_7-CO-\{$
Sebacic acid	Sebazoyl	$\{-CO-(CH_2)_8-CO-\{$

[a] *C*-Substitution products should be named systematically.

Table 23 (continued)

Name of the acid[a]	Group name	Formula
Unsaturated aliphatic carboxylic acids (non-limiting list)		
Acrylic acid	Acryloyl	$H_2C=CH-CO-$ }
Propiolic acid	Propioloyl	$HC\equiv C-CO-$ }
Methacrylic acid	Methacryloyl	$H_2C=C(CH_3)-CO-$ }
Crotonic acid (*trans*)	Crotonoyl	⎫ $H_3C-CH=CH-CO-$ }
Isocrotonic acid (*cis*)	Isocrotonoyl	⎭
Oleic acid	Oleoyl	$CH-(CH_2)_7-CH_3$ ⎫
		‖
Elaidic acid (*trans*)	Elaidoyl	$CH-(CH_2)_7-CO-$ } ⎭
Maleic acid (*cis*)	Maleoyl	⎫ $-CO-CH=CH-CO-$ }
Fumaric acid (*trans*)	Fumaroyl	⎭
Muconic acid	Muconoyl	$-CO-CH=CH-CH=CH-CO-$ }
Citraconic acid (*cis*)[a]	Citraconoyl	⎫ $-CO-C(CH_3)=CH-CO-$ }
Mesaconic acid (*trans*)[a]	Mesaconoyl	⎭
Cyclic carboxylic acids (non-limiting list)		
Camphoric acid	Camphoroyl	
Benzoic acid	Benzoyl	C_6H_5-CO- }
Phthalic acid (*o*)	Phthaloyl	⎫ $C_6H_4(CO-$ }$)_2$
(*Iso,Tere*)phthalic acid (*m,p*)	(*Iso,Tere*)phthaloyl	⎭
Naphthoic acid	Naphthoyl	$C_{10}H_7-CO-$ }
Toluic acid (*o,m,p*)	Toluoyl	$H_3C-C_6H_4-CO-$ }
Hydratropic acid	Hydratropoyl	$C_6H_5-CH(CH_3)-CO-$ }
Atropic acid	Atropoyl	$C_6H_5-C(=CH_2)-CO-$ }
Cinnamic acid	Cinnamoyl	$C_6H_5-CH=CH-CO-$ }
Nicotinic acid	Nicotinoyl	
Isonicotinic acid	Isonicotinoyl	

[a] *C*-Substitution products should be named systematically.

Table 23 (continued)

Name of the acid	Group name	Formula
Peroxy acids		
Performic acid	–	HC(O)OOH
Peracetic acid	–	H₃C–C(O)OOH
Perbenzoic acid	–	C₆H₅–C(O)OOH

Table 24. Hydroxy and alkoxycarboxylic acids (non-limiting list)

Name of the acid	Group name	Formula
Glycolic acid	Glykoloyl	$HO-CH_2-CO-\}$
Lactic acid	Lactoyl	$H_3C-CH(OH)-CO-\}$
Glyceric acid	Glyceroyl	$HO-CH_2-CH(OH)-CO-\}$
Tartronic acid	Tartronoyl	$\{-CO-CH(OH)-CO-\}$
Malic acid	Maloyl	$\{-CO-CH_2-CH(OH)-CO-\}$
Tropic acid	Tropoyl	$C_6H_5-CH(CH_2OH)-CO-\}$
Benzilic acid	Benziloyl	$(C_6H_5)_2C(OH)-CO-\}$
Mandelic acid	Mandeloyl	$C_6H_5-CH(OH)-CO-\}$
Salicylic acid	Salicyloyl	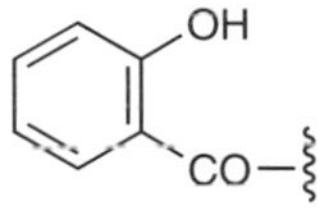
Anisic acid	Anisoyl	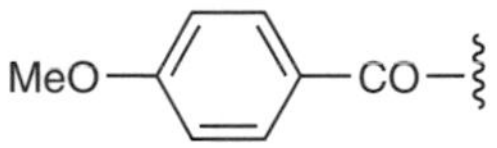
Vanillic acid	Vanilloyl	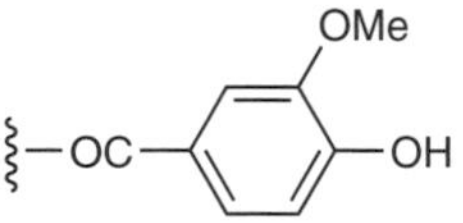
Veratric acid	Veratroyl	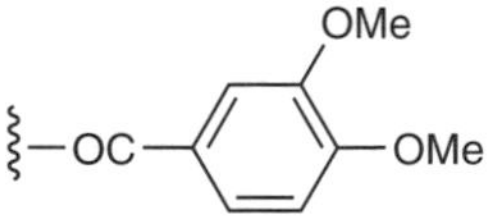
Piperonylic acid	Piperonyloyl	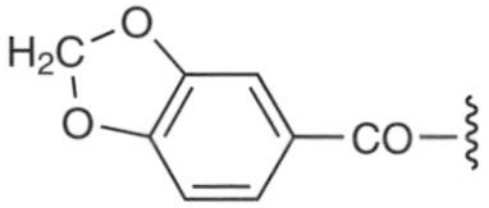

Table 24 (continued)

Name of the acid	Group name	Formula
Protocatechuic acid	Protocatechuoyl	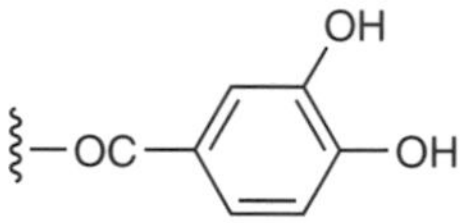
Gallic acid	Galloyl	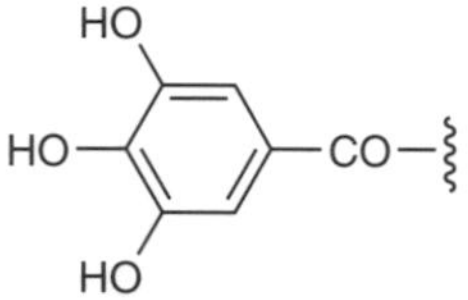
Citric acid	–	$\xi-CO-CH_2-C(OH)-CH_2-CO-\xi$ with $CO-\xi$ branch
Mevalonic acid	–	$HO-CH_2-CH_2-C(Me)-CH_2-CO-\xi$ with OH branch

Table 25. Oxocarboxylic acids

Name of the acid	Group name	Formula
Glyoxylic acid	Glyoxyloyl	$OHC-CO-\xi$
Pyruvic acid	Pyruvoyl	$H_3C-CO-CO-\xi$
Acetoacetic acid	Acetoacetyl	$H_3C-CO-CH_2-CO-\xi$
Levulinic acid	–	$H_3C-CO-CH_2-CH_2-CO-\xi$
Mesoxalic acid	Mesoxalyl	$\xi-CO-CO-CO-\xi$
	Mesoxalo	$HOOC-CO-CO-\xi$
Oxalacetic acid	Oxalacetyl	$\xi-CO-CH_2-CO-CO-\xi$
	Oxalaceto	$HOOC-CO-CH_2-CO-\xi$

Table 26. The most important α-aminocarboxylic acids (amino acids)

Amino acid	Abbrev.[a]		Group name	Formula
Alanine	Ala	(A)	Alanyl	$H_3C-CH(NH_2)-CO-\xi$
β-Alanine			β-Alanyl	$H_2N-CH_2-CH_2-CO-\xi$
Arginine	Arg	(R)	Arginyl	$HN=C-NH-(CH_2)_3-CH-CO-\xi$ (with NH_2 on C and NH_2 on CH)
Cystathionine	–		Cystathionyl	$S\langle CH_2-CH(NH_2)-CO-\xi ;\ CH_2-CH_2-CH(NH_2)-CO-\xi$
Cysteine	Cys	(C)	Cysteinyl	$HS-CH_2-CH(NH_2)-CO-\xi$
Cystine	–		Cystyl	$S-CH_2-CH(NH_2)-CO-\xi$ / $S-CH_2-CH(NH_2)-CO-\xi$
Glycine	Gly	(G)	Glycyl	$H_2N-CH_2-CO-\xi$
Histidine	His	(H)	Histidyl	$HC=C-CH_2-CH(NH_2)-CO-\xi$ with imidazole ring $HN{-}\overset{\text{C}}{\underset{H}{}}{-}N$
Homocysteine	–		Homocysteinyl	$HS-CH_2-CH_2-CH(NH_2)-CO-\xi$
Homoserine	–		Homoseryl	$HO-CH_2-CH_2-CH(NH_2)-CO-\xi$
Isoleucine	Ile	(I)	Isoleucyl	$C_2H_5-CH(CH_3)-CH(NH_2)-CO-\xi$
Lanthionine	–		Lanthionyl	$S\langle CH_2-CH(NH_2)-CO-\xi ;\ CH_2-CH(NH_2)-CO-\xi$
Leucine	Leu	(L)	Leucyl	$(CH_3)_2CH-CH_2-CH(NH_2)-CO-\xi$
Lysine	Lys	(K)	Lysyl	$H_2N-(CH_2)_4-CH(NH_2)-CO-\xi$
Methionine	Met	(M)	Methionyl	$H_3C-S-CH_2-CH_2-CH(NH_2)-CO-\xi$
Norleucine	–		Norleucyl	$H_3C-(CH_2)_3-CH(NH_2)-CO-\xi$
Norvaline	–		Norvalyl	$H_3C-(CH_2)_2-CH(NH_2)-CO-\xi$
Ornithine	–		Ornithyl	$H_2N-(CH_2)_3-CH(NH_2)-CO-\xi$
Phenylalanine	Phe	(F)	Phenylalanyl	$C_6H_5-CH_2-CH(NH_2)-CO-\xi$
Proline	Pro	(P)	Prolyl	$H_2C-CH-CO-\xi$ / $H_2C{-}\overset{\text{C}}{\underset{H_2}{}}{-}NH$
Sarcosine	–		Sarcosyl	$H_3C-NH-CH_2-CO-\xi$

Table 26 (continued)

Amino acid	Abbrev.[a]		Group name	Formula
Serine	Ser	(S)	Seryl	$HO-CH_2-CH(NH_2)-CO-\}$
Threonine	Thr	(T)	Threonyl	$H_3C-CH(OH)-CH(NH_2)-CO-\}$
Tryptophan	Trp	(W)	Tryptophyl	
Thyronine	–		Thyronyl	
Tyrosine	Tyr	(Y)	Tyrosyl	
Valine	Val	(V)	Valyl	$(H_3C)_2CH-CH(NH_2)-CO-\}$
Aspartic acid*	Asp	(D)[b]	α-Aspartyl	$HOOC-CH_2-CH(NH_2)-CO-\}$
			β-Aspartyl	$\{-CO-CH_2-CH(NH_2)-COOH$
			Aspartoyl	$\{-CO-CH_2-CH(NH_2)-CO-\}$
Asparagine	Asn	(N)[b]	Asparaginyl	$H_2N-CO-CH_2-CH(NH_2)-CO-\}$
Glutamic acid	Glu	(E)[c]	α-Glutamyl	$HOOC-CH_2-CH_2-CH(NH_2)-CO-\}$
			γ-Glutamyl	$\{-CO-CH_2-CH_2-CH(NH_2)-COOH$
			Glutamoyl	$\{-CO-CH_2-CH_2-CH(NH_2)-CO-\}$
Glutamine	Gln	(Q)[c]	Glutaminyl	$H_2N-CO-CH_2-CH_2-CH(NH_2)-CO-\}$

[a] IUPAC-IUB Commission on Biochemical Nomenclature, Pure Appl. Chem. **1972**, *31*, 641. The three-letter symbols are meant for normal use; the one-letter symbols should only be applied for special cases, above all when space is restricted.

[b] In addition, it is recommended to use the letter B when it is not clear wether **asparagine** or **aspartic acid** is present.

[c] The letter Z is used when it is not clear whether **glutamine** or **glutamic acid** is present. Each other unknown amino acid is characterized by the letter X.

Other amino acids and amidic acids

Hippuric acid	?	$C_6H_5-CO-NH-CH_2-COOH$
Anthranilic acid	?	

Table 26 (continued)

Amino acid	Group name	Formula
Carbamic acid	Carbamoyl	$H_2N-CO-\xi$
Carbazic acid	Carbazoyl	$H_2N-NH-CO-\xi$
Allophanic acid	Allophanoyl	$H_2N-CO-NH-CO-\xi$
Hydantoic acid	–	$H_2N-CO-NH-CH_2-COOH$

Table 27. Lactones and lactams

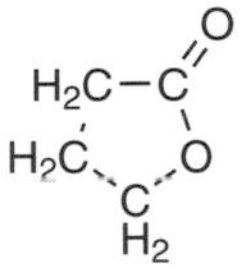

γ-Butyrolactone

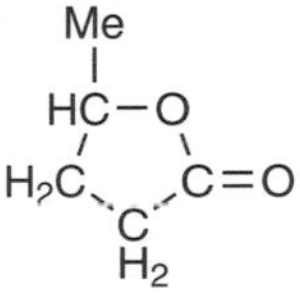

γ-Valerolactone

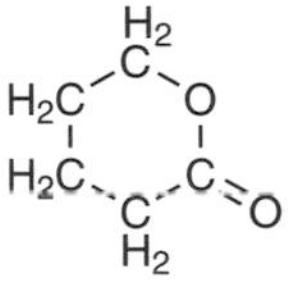

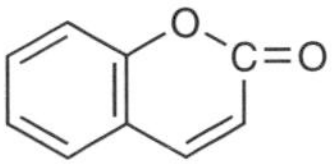

δ-Valerolactone

Coumarin

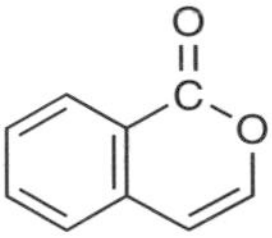

Isocoumarin

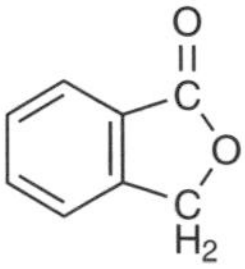

Phthalide

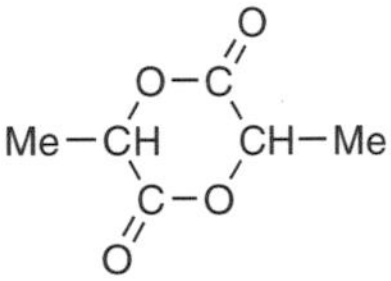

Dilactide

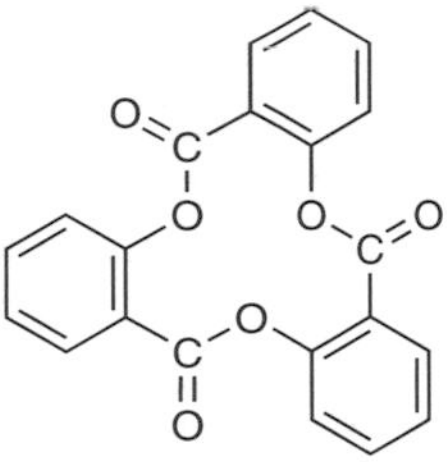

Trisalicylide

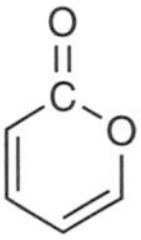

2-Pyrone or α-Pyrone

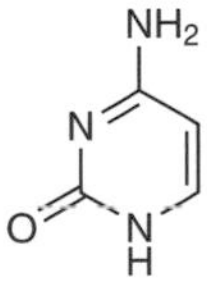

Cytosine(C)[a]

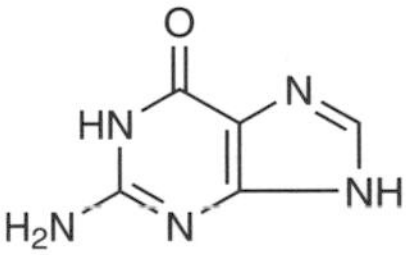

Guanine(G)[a]

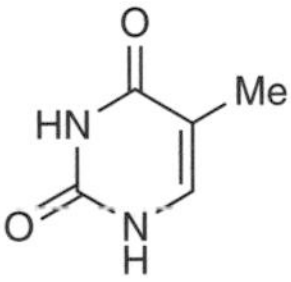

Thymine(T)[a]

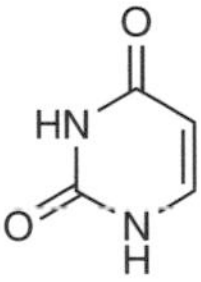

Uracil(U)[a]

[a] Letter symbols as used for representations of nucleic acids.

Table 28. Sulfonic acids and corresponding substituent groups

Sulfonic acid	Group name	Formula
Sulfanilic acid	–	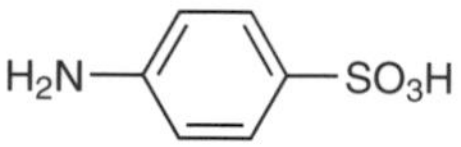
Naphthionic acid	–	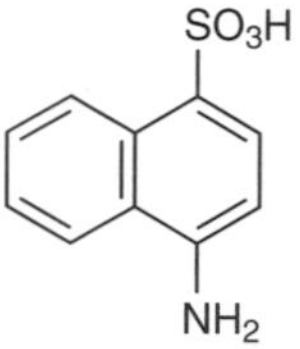
Taurine	Tauryl	$H_2N-CH_2-CH_2-SO_2-$ ⧵
Methanesulfonic acid	Mesyl	H_3C-SO_2- ⧵
p-Toluenesulfonic acid	Tosyl	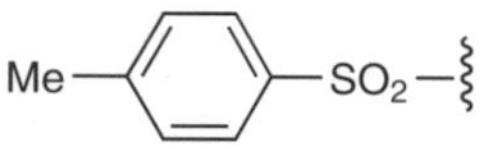

Table 29. Aldehydes

Aldehyde	Formula
Formaldehyde	HCHO
Acetaldehyde	H_3C-CHO
Propionaldehyde	H_3C-CH_2-CHO
Butyraldehyde	$H_3C-CH_2-CH_2-CHO$
Isobutyraldehyde	$(H_3C)_2CH-CHO$
Valeraldehyde	$H_3C-CH_2-CH_2-CH_2-CHO$
Isovaleraldehyde	$(H_3C)_2CH-CH_2-CHO$
Acrylaldehyde or Acrolein	$H_2C=CH-CHO$
Crotonaldehyde	$H_3C-CH=CH-CHO$
Benzaldehyde	C_6H_5-CHO
Cinnamaldehyd	$C_6H_5-CH=CH-CHO$
p-Anisaldehyde	
Nicotinaldehyde	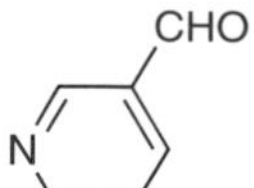

Table 29 (continued)

Aldehyde	Formula

2-Furaldehyde or Furfural

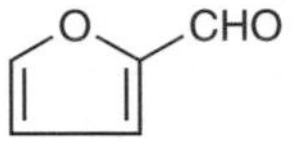

Glyceraldehyde	HO–CH$_2$–CH(OH)–CHO
Glycolaldehyde	HO–CH$_2$–CHO
Citral	(H$_3$C)$_2$C=CH–(CH$_2$)$_2$–C(CH$_3$)=CH–CHO

Vanillin

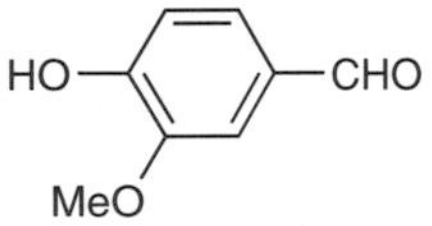

Piperonal

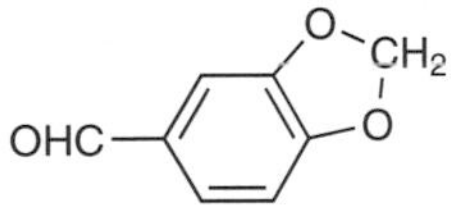

Glyoxal	OHC–CHO
Malonaldehyde	OHC–CH$_2$–CHO
Succinaldehyde	OHC–(CH$_2$)$_2$–CHO
Glutaraldehyde	OHC–(CH$_2$)$_3$–CHO
Adipaldehyde	OHC–(CH$_2$)$_4$–CHO

Phthalaldehyde

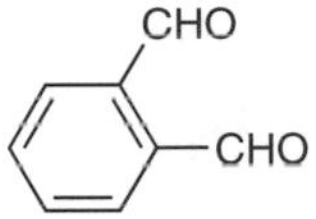

Isophthalaldehyde

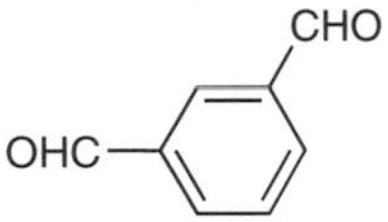

Terephthalaldehyde

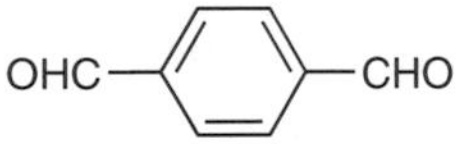

Table 30. Ketones

Ketone	Formula
Acetone	$H_3C-CO-CH_3$
Propiophenone	$C_6H_5-CO-CH_2-CH_3$
Desoxybenzoin	$C_6H_5-CH_2-CO-C_6H_5$
Chalcone	$C_6H_5-CH=CH-CO-C_6H_5$
Biacetyl	$H_3C-CO-CO-CH_3$
Benzil	$C_6H_5-CO-CO-C_6H_5$

2,2'-Furil

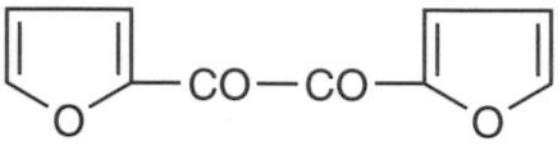

2,2'-Furoin

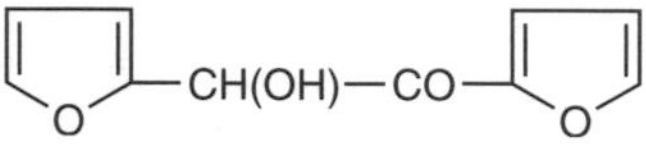

Acetoin	$H_3C-CH(OH)-CO-CH_3$
Benzoin	$C_6H_5-CH(OH)-CO-C_6H_5$

Anthrone

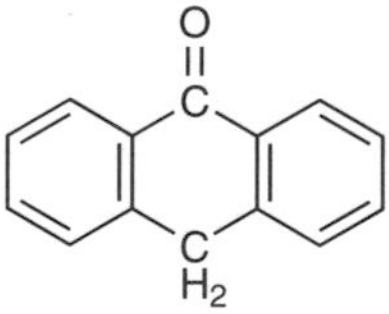

Phenanthrone

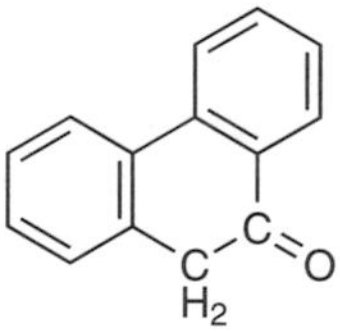

Ketone substituent groups

Acetonyl	$H_3C-CO-CH_2-$
Acetonylidene	$H_3C-CO-CH=$
Phenacyl	$C_6H_5-CO-CH_2-$
Phenacylidene	$C_6H_5-CO-CH=$

Table 31. Alcohols

Alcohol	Formula
Allyl alcohol	$H_2C=CH-CH_2-OH$
tert-Butyl alcohol	$(H_3C)_3C-OH$
Benzyl alcohol	$C_6H_5-CH_2-OH$
Phenethyl alcohol	$C_6H_5-CH_2-CH_2-OH$

Salicyl alcohol

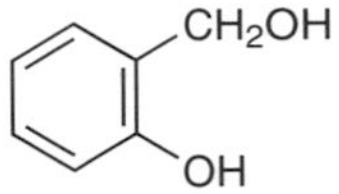

Alcohol	Formula
Crotyl alcohol	$H_3C-CH=CH-CH_2-OH$

Geraniol

$$(CH_3)_2C=CH-CH_2-CH_2-\underset{\underset{CH_3}{|}}{C}=CH-CH_2-OH$$

Farncsol

$$(CH_3)_2C=CH-CH_2-CH_2-\underset{\underset{CH}{||}}{C}-CH_3$$
$$HO-H_2C-CH=\underset{\underset{CH_3}{|}}{C}-CH_2-CH_2-CH$$

Phytol

$$(CH_3)_2CH-(CH_2)_3-\underset{\underset{CH_3}{|}}{\overset{\overset{CH_3}{|}}{CH}}-CH_2-CH_2$$
$$HO-H_2C-CH=\underset{\underset{CH_3}{|}}{C}-(CH_2)_3-\underset{\underset{CH_3}{|}}{CH}-CH_2$$

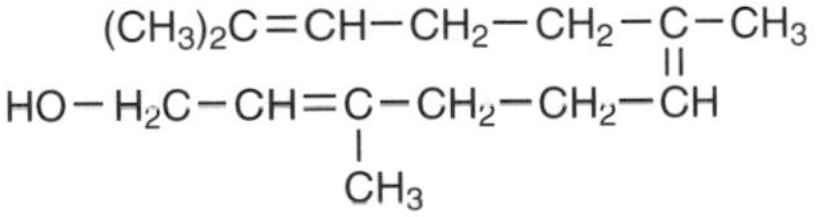

Alcohol	Formula
Ethylene glycol	$HO-H_2C-CH_2-OH$
Propylene glycol	$H_3C-CH(OH)-CH_2-OH$
Glycerol	$HO-H_2C-CH(OH)-CH_2-OH$
Pinacol	$(H_3C)_2C(OH)-C(OH)(CH_3)_2$
Erythritol	$HO-H_2C-CH(OH)-CH(OH)-CH_2-OH$
Pentaerythritol	$C(CH_2-OH)_4$

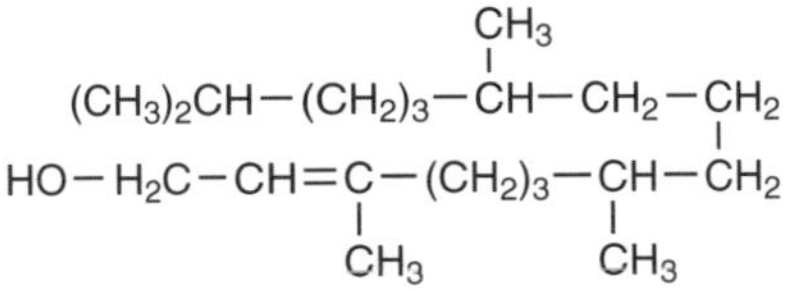

Menthol

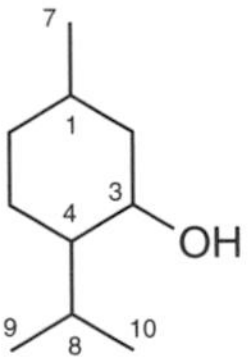

Borneol

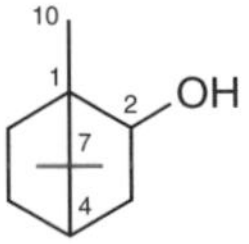

Table 32. Phenols

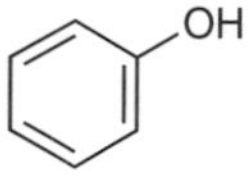

Phenol

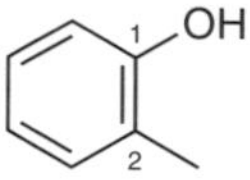

o-Cresol

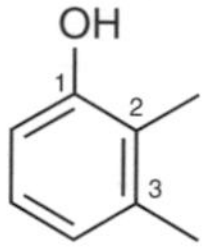

2,3-Xylenol

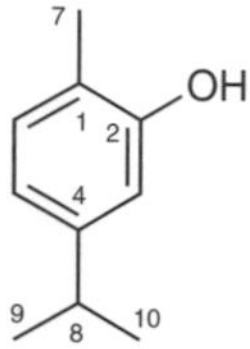

Carvacrol

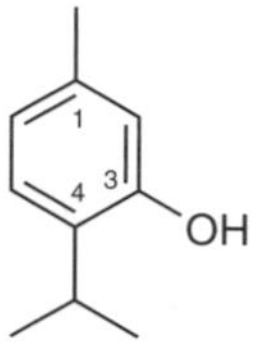

Thymol

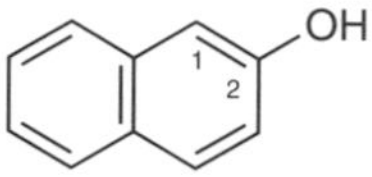

2-Naphthol

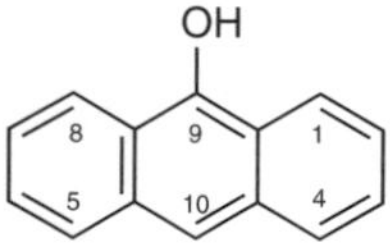

9-Anthrol

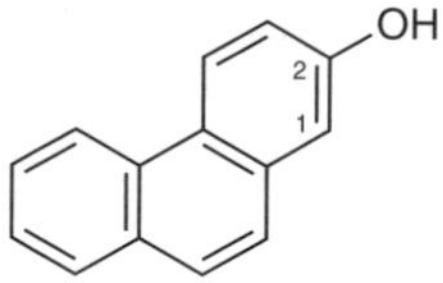

2-Phenanthrol

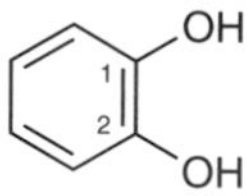

Pyrocatechol

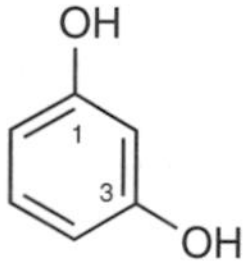

Resorcinol

Hydroquinone

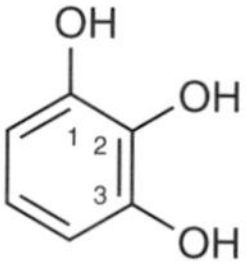

Pyrogallol

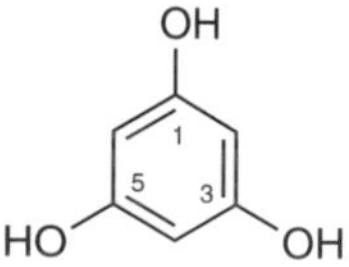

Phloroglucinol

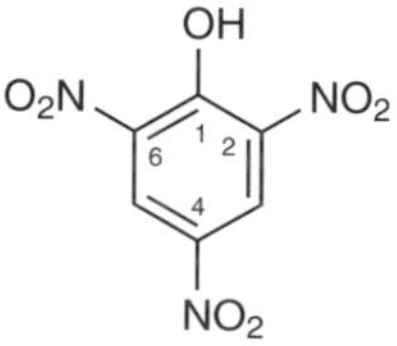

Picric acid

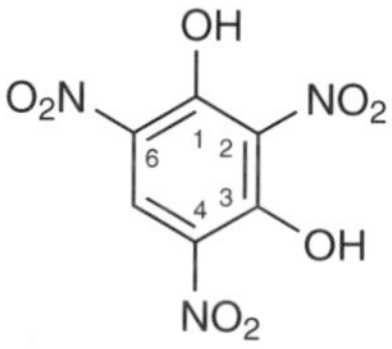

Styphnic acid

Table 33. Ethers

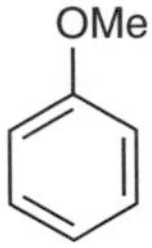

Anisole

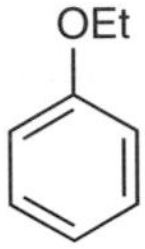

Phenetole

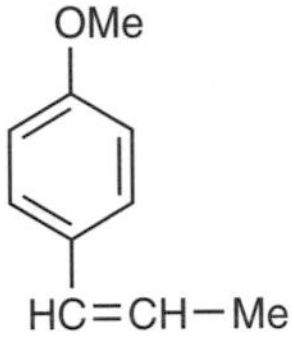

Anethole

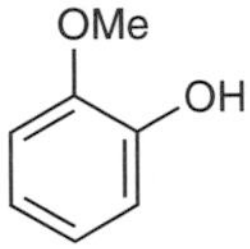

Guaiacol

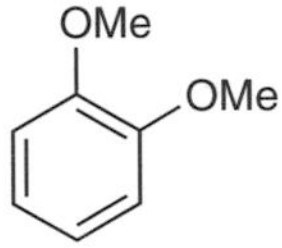

Veratrole

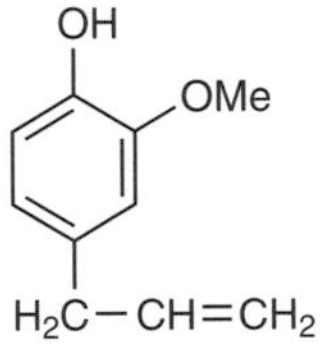

Eugenol

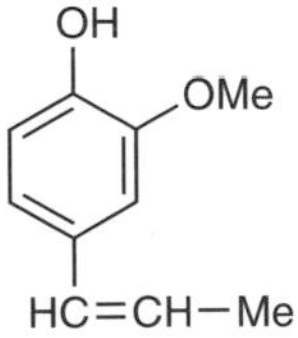

Isoeugenol

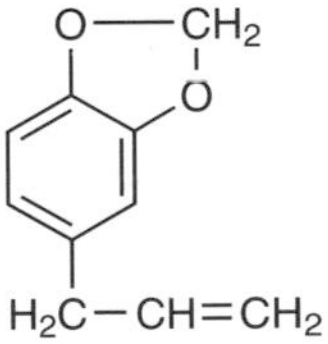

Safrole

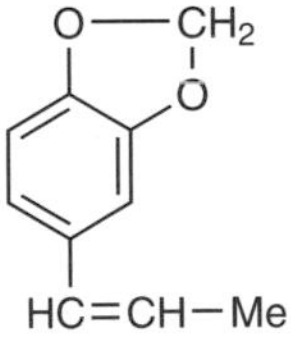

Isosafrole

Table 34. Amines, ammonium compounds, amides

Trivial name	Formula
Aniline	$C_6H_5–NH_2$
o,m,p-Anisidine	$H_3C–O–C_6H_4–NH_2$
o,m,p-Phenetidine	$C_2H_5–O–C_6H_4–NH_2$
o,m,p-Toluidine	$H_3C–C_6H_4–NH_2$
2,4-Xylidine	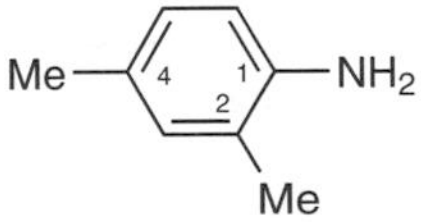
Adenine (A)[a]	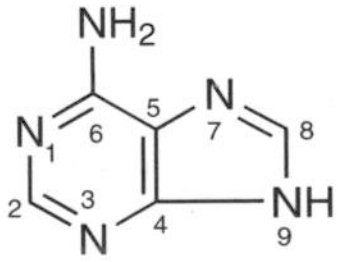

[a] Letter symbol as used in representations of nucleic acids.

Table 34 (continued)

Trivial name	Formula
Colamine	$HO-CH_2-CH_2-NH_2$
Sphingosine	$H_3C-(CH_2)_{12}-CH=CH-CH-CH-CH_2-OH$ with OH and NH_2 substituents
Ephedrine	$Ph-CH-CH-CH_3$ with OH and $NHMe$ substituents
Benzedrine	$Ph-CH_2-CH-CH_3$ with NH_2 substituent
Adrenaline	
Putrescine	$H_2N-CH_2-CH_2-CH_2-CH_2-NH_2$
Cadaverine	$H_2N-CH_2-CH_2-CH_2-CH_2-CH_2-NH_2$
Benzidine	
Choline (bromide etc.)	$HO-CH_2-CH_2-\overset{\oplus}{N}(CH_3)_3Br^{\ominus}$
Betaine	$^{\ominus}OOC-CH_2-\overset{\oplus}{N}(CH_3)_3$
Betaine (hydrobromide etc.)	$HOOC-CH_2-\overset{\oplus}{N}(CH_3)_3Br^{\ominus}$ etc.
Di, Triacetamide	$HN(CO-CH_3)_2, N(CO-CH_3)_3$
Di, Tribenzamide	$HN(CO-C_6H_5)_2, N(CO-C_6H_5)_3$

Table 35. Halogen compounds

Trivial name	Formula
Methylenechloride, -bromide, -iodide, -fluoride	H_2CX_2
Fluoro-, chloro-, bromo-, iodoform	HCX_3
Tetra-fluoro-, -iodo-, -bromo-, -chlorocarbon	CX_4
Benzal chloride, -bromide, -iodide, -fluoride	$C_6H_5-CHX_2$
Benzotrichloride, -bromide, -iodide, -fluoride	$C_6H_5-CX_3$
Di-iodo-, -bromo-, -chloro-, -fluoro-carbene (-methylene)	$:CX_2$
Phosgene or carbonyl dichloride etc.	$Cl-CO-Cl$
Thiophosgene or Thiocarbonyl dichloride etc.	$Cl-CS-Cl$

Subject Index